Environmental Footprints and Eco-design of Products and Processes

Series editor

Subramanian Senthilkannan Muthu, SGS Hong Kong Limited, Hong Kong, Hong Kong SAR

More information about this series at http://www.springer.com/series/13340

Ali Alsamawi · Darian McBain
Joy Murray · Manfred Lenzen
Kirsten S. Wiebe

The Social Footprints of Global Trade

Springer

Ali Alsamawi
School of Physics
University of Sydney
Sydney, NSW
Australia

Darian McBain
Sustainable Development
Thai Union Group
Bangkok
Thailand

Joy Murray
School of Physics
University of Sydney
Sydney, NSW
Australia

Manfred Lenzen
School of Physics
University of Sydney
Sydney, NSW
Australia

Kirsten S. Wiebe
Department of Energy and Process Engineering
Norwegian University of Science and Technology
Trondheim
Norway

ISSN 2345-7651 ISSN 2345-766X (electronic)
Environmental Footprints and Eco-design of Products and Processes
ISBN 978-981-10-4135-8 ISBN 978-981-10-4137-2 (eBook)
DOI 10.1007/978-981-10-4137-2

Library of Congress Control Number: 2017935552

Printed on acid-free paper

This Springer imprint is published by Springer Nature
The registered company is Springer Nature Singapore Pte Ltd.
The registered company address is: 152 Beach Road, #21-01/04 Gateway East, Singapore 189721, Singapore

Acknowledgements

We would like to acknowledge the support and assistance provided to us by our colleagues from the Integrated Sustainability Analysis Group, School of Physics, University of Sydney.

Contents

1 **Introduction** 1
1.1 Importance of International Trade 1
1.2 Globalization 2
1.3 Social Indicators in Global Supply Chains 3
References 5

2 **Methodology** 7
2.1 Introduction to Input Output Analysis 7
2.2 Satellite Account 8
2.3 Extended to Include Social Indicators 9
References 12

3 **Introduction to the Social Footprints of Global Trade** 13
3.1 Measuring Social Impacts in Supply Chains 14
References 16

4 **Review of Social Accounting Methodologies** 19
4.1 Supplier Social Audits 19
4.2 Third Party Certification and Verification 19
4.3 Purchasing and Procurement Practices 20
4.4 Multi-regional Input–Output Analysis 20
References 23

5 **Review of Social Metrics and Social Footprinting** 27
5.1 Introduction: Social Indicators and Metrics 27
5.2 Footprinting History and Methodology 30
References 33

6 **A Social Footprint of Nations: A Comparative Study of the Social Impact of Work** 35
6.1 Introduction 35
6.2 Background 37

6.3 Methodology 38
6.4 Limitations of the MRIO Methodology for Footprinting. 39
6.5 Results 40
6.6 Discussion 46
6.6.1 Role of Governments 46
6.6.2 Role of Business and Institutional Investors 47
6.6.3 Role of Consumers 48
6.7 Conclusion 48
References. 50

7 Case Study—Assessing Social Impacts in the Seafood Industry. . . . 53
7.1 Introduction 53
7.2 Is Thailand the Highest Risk?. 54
7.3 Key Risk Factors Around Modern Slavery and Human Rights in the Seafood Sector 56
7.4 The Unanswered Question—Who Pays for Improved Working Conditions?. 57
7.5 Conclusion 58

8 Raising the International Poverty Line—A Comparison of Necessary Adjustments of Final Demand Spending in OECD and Non-OECD Countries 59
8.1 Introduction 59
8.2 Population, Employment, Working Below the Poverty Line and Final Demand 60
8.3 "Working Below the Poverty Line" Footprint 61
8.4 Policy Scenarios: Raising Workers' Wages to Poverty Line. 65
8.5 Discussion 66
References. 67

9 The Inequality Footprints of Nations; A Novel Approach to Quantitative Accounting of Income Inequality 69
9.1 Introduction 69
9.2 Background 70
9.2.1 Inequality as a Global Issue 70
9.2.2 Within Country Inequality. 71
9.2.3 Responsibility 73
9.3 Methodology and Data Sources 74
9.3.1 Data Sources. 74
9.3.2 Basic Input-Output Theory 74
9.3.3 Extended Input-Output Analysis 75
9.3.4 Inequality Satellite Accounts. 76
9.3.5 Inequality Footprints. 79
9.3.6 Qualifications 81

9.4 Results 81
9.5 Conclusion 87
References. 89

10 Conclusion 93
References. 94

Appendix A: List of Countries Covered in This Paper 95

Appendix B: Countries and Years for Which Data Are Available 101

Appendix C 113

Appendix D: Further Comparisons of Lorenz Curves Based on Measured Income and Employment Distributions and Fitted Based on Gini Index Data 117

Appendix E 119

Appendix F 121

References 129

Summary

Globalization and growth in emerging economies in the past decades have driven an increase in the volume of international trade. This means that the number of workers in a country working to supply the consumption of other countries has increased. Concerns about the conditions of the workers in emerging economies increased after, for instance, the discovery of child labour and human trafficking in the Thai fishing sector and the collapse of the Rana Plaza building in Bangladesh. Most of the workers killed in that collapse were working to produce textiles consumed in European countries. There is a growing interest in the past years by decision-makers and international organizations in issues around the social impact of supply chains. For instance, in early 2016 the International Labour Organization published a series of papers about recent work in global supply chains and organized an international meeting to discuss the issue.

This book investigates a number of social issues that have been linked to a Multi-Regional Input-Output tables (MRIO) to measure what are called *Social Footprints of Nations*. Chapter 1 provides a brief introduction to this work. Chapter 2 presents some principles of input-output analysis. Chapter 3 presents an introduction to social footprints. Chapter 4 contrasts the MRIO approach with other popular methods of social supply chain analysis. Chapter 5 provides a review of social metrics and social footprints with some examples of social indicators. In Chap. 6, we investigate eight social indicators and develop a method to measure the social footprint of nations by compiling those indicators for more than 135 countries. Indicators include employment, income and days lost due to accidents. We compare these data for the average worker across developed and developing nations. Chapter 7 presents a new term—*Poverty Footprint*—which includes all workers working in global supply chains and earning less than $1.25 a day. This chapter shows that the OECD countries would only need to spend about 0.36% more on final demand to ensure all workers' earnings are at least as high as the international poverty line. Chapter 8 presents a case study of the *Inequality Footprint*. An inequality footprint shows the link that each country's domestic economic activity has to income distribution elsewhere in the world. In this chapter, we use economic input-output analysis to calculate the inequality footprint of nations. We use employment and household income accounts for 187 countries and a historical time series dating back to 1990. Chapter 9 presents a case study on assessing the social impact of Thailand's seafood industry.

This book presents a new term: *Social Footprints*. Social footprints can play an important role in illustrating the impact of international trade on social indicators within countries. Thus, the results presented in this study are crucial to policy formulation that targets labour rights and working conditions in the local, regional and global context.

Chapter 1
Introduction

1.1 Importance of International Trade

Two hundred years ago, David Ricardo developed his theory of Comparative Advantage which assumed that international trade led to economic development, growth and prosperity (Ricardo 1817). He explained that international trade and exchanging commodities across national borders are benefits for countries even if they produce the same product. In his theory, he suggests that a country could specialize to produce a product that is internationally competitive and could trade with other nations for goods that are too costly to produce domestically. For instance, no doubt it is better for Finland to import bananas from other nations than grow them nationally. As the world becomes more globalized through the technology revolution in the 21st century, people now are keen to experience other nations' food, drink, and culture. For example international trade allows the British to taste an Indian tea, and Americans to drink French wine. But how does Comparative Advantage theory work?

Suppose that Saudi Arabia produces 10 thousand metric tons of milk, which in turn require a certain amount of water in the production process. While Saudi Arabia is an arid land with only a few unsustainable aquifers, it relies on desalination to fill the shortage of water for municipal and agricultural purposes. Thus the 10 thousand metric tons of milk produced in Saudi Arabia may require, for example, one million barrels of crude oil in production (for desalination). This 'one million barrels of crude oil' could be used instead to produce other products (for example 50 thousand metric tons of chemical fertilizers) in Saudi Arabia. Another country, India, one of the world's major producers of milk with low cost production, could produce 25 thousand metric tons of milk with the one million barrels of crude oil. Since India has fewer natural resources in terms of crude oil than Saudi Arabia, only 30 thousand metric tons of chemical fertilizers could be produced with the one million barrels of crude oil. From this point, there is a comparative

A. Alsamawi et al., *The Social Footprints of Global Trade*,
Environmental Footprints and Eco-design of Products and Processes,
DOI 10.1007/978-981-10-4137-2_1

Table 1.1 The advantage in changing commodities between India and Saudi Arabia (hypothetical data)

	Milk	Chemical fertilizers
Saudi Arabia	10	50
India	25	30
Advantages	15	20

advantage for Saudi Arabia to produce chemical fertilizers for India and India to produce milk for Saudi Arabia (Table 1.1).

1.2 Globalization

Recently, Ricardo's theory has been significantly promoted by modern globalization, which encourages countries to exchange their resources and commodities across their national borders. Rapid growth, and changes from agricultural to industrialized economies, in most Asian countries have also played an important role to increase their trade with Europe and US. According to Daly and Farley 2011, globalization is becoming a "*panacea for economic problems*". International trade could play a valuable role in improving workers' conditions because, for instance, in 2008 about $16 trillion (equivalent to 30% of the world's produced goods and services) worth of products were sold in exports and imports processes (Krugman et al. 2012, p. 10). Meanwhile, there are huge numbers of workers worldwide working to produce products consumed elsewhere (Alsamawi et al. 2014). As an example of that, in 2010, about 70% of the domestic workforce in Madagascar and Papua New Guinea was working for the sake of exports. However, there is a question we could ask: Did open borders for free trade and increased liberalization in the past decade between Africa and Europe help African workers out of poverty or improve their working conditions? Although globalization can create new jobs through trade and investments, it usually calls for new innovation and technologies that need highly skilled people, which in turn leads to the detriment of unskilled workers in both developing and developed countries. Today, with this rapid growth, corporations have enormously increased their *profit* through international trade. But to evaluate the performance of (for instance) a corporation, it's necessary to address the three dimensions of sustainability (*people*, *planet* and *profit*). The well-known definition of sustainable development according to (WCED 1987), is

> Development that meets the needs of the present without compromising the ability of future generations to meet their own needs.

Thus the question could be asked: What do we need to do to address the *people* and *planet* dimensions of sustainability?

1.3 Social Indicators in Global Supply Chains

As we can see, neither Ricardo's theory nor globalization mentioned the social and environmental aspects beyond the trade. Environmental and economic indicators are well known and have been extensively used by policy-makers, governments, academia and NGOs (more information about this in the methodology chapter). However, neglecting the third dimension (social or *people*) may have negative impacts on sustainability. The impact of international trade on social indicators is either neglected or rare to find. So the questions we can ask are: Is the trade that Ricardo and globalization are calling for fair? Do workers earn enough to survive?

Everyday, we are consuming and producing goods and services. Choosing best quality and the cheapest product are our main concerns when we make our daily purchases. But today the world has become more complex even in the case of making unsophisticated purchases and it's hard to choose one product among others with the same quality and price. Some of these products have labels, such as *fair trade*, which inform us about the impact of consuming a product to the ecosystem (environmental labeling) and/or to the working conditions of workers (social labeling). Thus, in addition to fulfilling our consumption appetite, consuming products that have fair trade labeling has benefits in social and environmental terms. Therefore, our choices can have a huge influence on our transformation into a sustainable society. Recently, fair trade has become a useful mechanism to inform consumers about the conditions under which the products were produced. In 1997, an international non-government organization, Global Reporting Initiative (GRI), was created by the United Nations, UNEP and Tellus institute to help corporations, governments and others to understand their impact on critical issues such as climate change and human rights along their supply chains.

Addressing working conditions in developing countries is not only essential for those countries but also developed nations want to know what's happening in developing nations. Concerns about the conditions of work by developed countries have increased after some developing countries became themselves industrialized nations (for example China). The people of advanced nations consume the majority of products exported into supply chains snaking around the world. Thus, as they are responsible for their emissions embodied with the traded goods and services, developed nations may also be responsible in part for challenging working conditions in countries where consumed products were produced.

Some media reports and articles have recently revealed the conditions of people working for other countries. For instance, in 2012 a fire in a garment factory in Dhaka killed about 117 workers (Burke and Hammadi 2012). In the following year, 2013, another disaster occurred in the same country where an eight-story building collapsed (known as Rana Plaza) in Dhaka and killed about 1129 workers (Butler and Sarah 2013). We heard about these disasters because they had huge media coverage worldwide. Thus, this could lead us to expect that there are many other hidden accidents occurring each day as a result of international trade movement. Also, in terms of inequality, in early 2014 Oxfam revealed at the World Economic

Forum held in Davos, Switzerland that 85 people owned wealth equivalent to about half of the world population. In the following year, 2015, in the same forum it was revealed that 1% of the world's population would own more than all the rest in 2016. Moreover, according to Alsamawi et al. (2014), most developed countries (which enjoy low income inequality) import foods and services from countries that have high within country inequality.

Therefore, there is a pressing need to create robust satellite accounts that hold social data. Consequently, a Multi-Regional Input–Output (MRIO) table and satellite accounts have been used to calculate what is called *Social Footprints* (*Footprinting* has been used previously to measure the impact of trade on environmental and economic indicators, but in this case it is used for social indicators). Social Footprints of a country include the domestic cases within the country plus those embodied within its imported goods and services. For instance, in the case of employment footprints (see Chap. 3 for more information), it includes the domestic workforce within a country plus those embodied within the imported commodities to that country in the same fiscal year. Debate about social footprinting and implications for international trade of working conditions in developing countries, has significantly increased in the past years. It's likely that the debate increased after the United Nations Environment Programme (UNEP) released a significant article focusing on the Social Life Cycle Assessment of products (UNEP 2009). While the term *Social Footprint* is new, principles of its work are similar to those of Corporate Social Responsibility (CSR). The definition of CSR according to Van Marrewijk 2003 is:

> In general, corporate sustainability and CSR refer to company activities – voluntary by definition – demonstrating the inclusion of social and environmental concerns in business operations and in interactions with stakeholders.

Thus, CSR is a tool that helps corporations to measure their environmental, economic and social activities along global supply chains. Although it is a good tool, not all companies measure their impacts in their supply chains. This leads us again to the realisation that the whole picture is not complete. In this regard, a need for a new technique, method and approach is recognized. Consequently, creating a world social account that holds (for example) the full-time equivalent workers per country per economic sector is needed. Coupling this account with the IO table allows us to calculate the direct (within a country) and indirect (in a supply chain) employment. To track these convoluted supply chains and show up the hidden conditions of workers, input–output analysis and in particular footprinting is an ideal tool. Indicators that are incorporated in this work include income distribution, employment, wages, occupational health and safety (this includes fatal and nonfatal accidents and days and wages lost), strikes and lockouts, unemployment and poverty in the world economy.

The outputs of coupling social accounts with an IO table are *social footprints* (more information in the following chapters). These findings can provide information to decision makers to help in guiding them to good decisions. Also they show the hidden conditions of workers in a global supply chain from production to consumption indicating where to concentrate efforts in addressing poor conditions. The results could be useful for think tanks, academia and international organizations. The results could be incorporated with other environmental and economic indicators to build a Triple Bottom Line (TBL) of the world economy. In this regard, any new policies, deals and legislation affecting working conditions in developing countries need to take the size of footprints into account. Both producing and consuming countries need to stand together to participate in formulation of these policies and laws. International organizations that directly deal with working conditions (such as UN, World Bank, Global Reporting Initiative, and International Labour Organization) can positively impact on working conditions. In 2015, two such deals have been implemented: one by the UN—Paris's climate change deal to reduce greenhouse gases emission (Paris Agreement); and the other by the World Trade Organization (WTO) to abolish subsidies from agricultural products (Nairobi Package).

References

Alsamawi, A., Murray, J., & Lenzen, M. (2014). The employment footprints of nations uncovering master-servant relationships. *Journal of Industrial Ecology, 18*(1), 59–70.

Burke, J., & Hammadi, S. (2012, November 26). *Bangladesh textile factory fire leaves more than 100 dead. The Observer*. London: The Guardian.

Butler, Sarah (2013, June 23). Bangladeshi factory deaths spark action among high-street clothing chains. The Observer. London: The Guardian. Retrieved July 20, 2013.

Daly, H. E., & Farley, J. C. (2011). *Ecological economics: Principles and applications* (2rd ed.). Island Press Publisher.

Krugman, P., Obstfeld, M., Melitz, M. (2012). *International economics: Theory and policy* (9rd ed.). Addison-Wesley (Pearson).

Nairobi Package, Nairobi Ministerial Declaration. (2015, December 19). Ministerial conference, 10th, Nairobi.

Paris Agreement, United Nations Treaty Collection. (2016, May 15). Retrieved May 16, 2016.

Ricardo, D. (1817). *On the principles of political economy and taxation* (3rd ed.). London: John Murray, Albemarle-Street, 1821 (Originally published in 1817). Online: http://www.systemics.com/docs/ricardo/principle.html

Van Marrewijk, M. (2003). Concepts and definitions of CSR and corporate sustainability: between agency and communion. *Journal of Business Ethics, 44*, 95–105.

UNEP. (2009). Guidelines for social life cycle assessment of products, United Nations Environment Program, Paris SETAC Life Cycle Initiative United Nations Environment Programme ISBN: 978-92-807-3021-0.

World Commission on Environment and Development. (1987). *Our Common future: Report of the world commission on environment and development*. Oxford: Oxford University Press.

Chapter 2
Methodology

2.1 Introduction to Input Output Analysis

In late November 2015, Turkish militants shot down a Russian fighter jet when it was fighting Islamic State militants in Syria on the basis of violating Turkish airspace, according to Turkish authorities. Vladimir Putin, president of Russia, took a series of political and economic actions against Turkey. One of those actions was to stop buying Turkish vegetables, which forced the government to look for other markets. This had a dramatic effect on the Turkish agriculture sector, which as a direct consequence had to lay off some workers. The transportation sector was also affected as it transported commodities from farms to ports. Packaging of those products was similarly affected. Meanwhile, other markets, for example Morocco and Tunisia would have gained from that action because to satisfy its demand, Russia had to find alternative trading partners. The number of those employed in the Moroccan agriculture sector would have increased, as well as in those employed in transportation and other sectors. Increasing exports may encourage Morocco to expand or build new ports to facilitate the exporting process, which in turn would indirectly benefit the manufacturing and construction sectors. The question has to be asked, how can we evaluate and measure the spillover effects of Putin's actions, in particular those related to economic sanctions?

One of the most robust techniques that is used to measure the impact of international trade on economic indicators is Multi-Regional Input-Output (MRIO) analysis. The root of this technique dates back to the 1930s when Wassily Leontief developed his theory about input-output techniques. The first input-output table created was for the US economy in 1936 (Leontief 1936). In his research he showed how changes in one economic sector could have an influence on other sectors in the US economy. For this pioneering work, he won the Nobel Memorial Prize in Economic Sciences in 1973. Input-output tables show the flow of goods and services among sectors in an economy. It consists of three blocks; $N \times N$ intermediate matrix $\mathbf{T}$, $N \times M$ final demand $\mathbf{y}$ and $K \times N$ primary or value added $\mathbf{v}$ block (Fig. 2.1). The intermediate block holds

A. Alsamawi et al., *The Social Footprints of Global Trade*,
Environmental Footprints and Eco-design of Products and Processes,
DOI 10.1007/978-981-10-4137-2_2

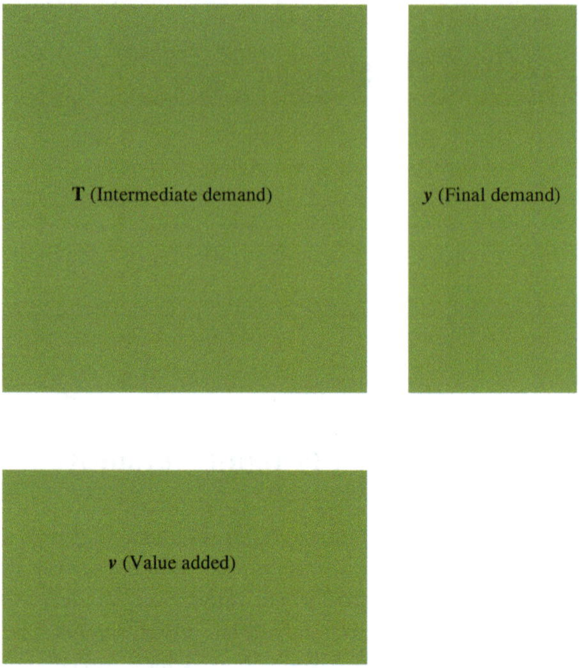

Fig. 2.1 Input output structure

monetary transactions from one sector to another within an economy. Final demand contains two main sub-blocks or categories: (1) *Gross National Expenditure* which includes households and government expenditure, gross fixed capital expenditure and change in inventories and (2) *Exports*. The value added or primary input block also contains two main sub-blocks or categories: (1) *Gross Domestic Products* which includes wages, gross operation surplus, taxes and subsidies (2) *Imports*.

The amount of sector i needed to produce one unit of sector j is called the technical coefficient aij. Thus, the technical coefficient matrix **A** that holds the entire interconnection among sectors in an economy shows the production recipes of those sectors, where $\mathbf{A} = (a_{ij}) = a_{ij}/x_j$, and x_j is the total output of sector j.

2.2 Satellite Account

The satellite account is an account that can be used to link physical data (e.g. tonnes) with monetary data that exist in the input output table. This account gives us more information about a country's national account data. While the IO table shows the interactions between sectors, households, government, and exports and imports, satellite accounts reveal information about the physical data of, for instance, the volume of emissions emitted from each sector of an economy. Satellite account units can be in tonnes, cubic meter, Full-Time Equivalent (FTE), etc. From recent years and on a regular basis, most developed and some developing countries have published data about satellite indicators.

Fig. 2.2 Sample of a concordance matrix

New classification

Raw data						
	1	0	0	0	0	0
	0	1	0	0	1	0
	0	0	1	1	0	0
	0	0	0	1	0	0
	0	0	0	0	1	1

The satellite account is a $1 \times N$ vector and has the same classification as **T** matrix in the MRIO table. In some MRIO, like Eora,[1] industries in certain countries have more details and are disaggregated into sub-sectors. From a first glance, disaggregation of the satellite account according to the **T** matrix seems an easy task to tackle. However, databases usually have lots of challenges. One of these challenges, for example, is that satellite data have different classifications than the input-output table's classifications. For instance, suppose you want to calculate a footprint of a country and the input-output table of that country has 17 classifications (for instance, International Standard Industrial Classification of All Economic Activities ISIC, Rev.3). Assume the available raw data for the satellite account has only three sectors (Agriculture, Manufacturing and Services). To overcome this problem, you build what is called a concordance matrix that has 0 and 1 values. The value 1 is assigned where a sector in the raw data corresponds to those in the MRIO classification, and 0 where it does not (Fig. 2.2).

One of the well-known uses of the satellite accounts and especially those dealing with environmental indicators is to measure a footprint of a country. The *footprint* is the sum of, for instance, the emissions that are associated with products consumed within a country plus those embodied within the imported goods and services from the rest of the world, subtracted from the emissions associated with the exports of that country. In recent years, this footprint technique has been developed to be used for carbon (Hertwich and Peters 2009), material (Wiedmann et al. 2013), net primary production (Haberl et al. 2007), and others. However, the third pillar of sustainability (social) seems neglected. In this regard, here I have presented a novel approach to measuring the impact of international trade on social indicators within countries by calculating the footprints of nations.

2.3 Extended to Include Social Indicators

To establish a healthy society or a corporation, the three vehicles of sustainability (environmental, social and economic) need to be moved in parallel. While the social dimension has been neglected in the past, an effort by academia and researchers in

[1]See www.worldmrio.com.

international organizations to investigate the social aspect is now underway. However, thus far few attempts have been added to the literature. These include, for example; employment and wages (Alsamawi et al. 2014), inequality (see Chap. 9), occupational health and safety (Simas et al. 2014; Alsamawi et al. 2016), and child labour (Gomez-Paredes et al. 2016). In each of these social cases and others (e.g. environmental and economic indicators), the method used to measure the footprints of nations follows the same process. In this work, I will illustrate the footprints of employment, income, poverty and occupational health and safety of nations. As I mentioned before, a Multi-Regional Input-Output table will be coupled with a physical satellite account to calculate the footprints using Leontief's inverse technique. To make it explicit, suppose that the IO table has three sectors in a country and a satellite account **Q** holds the total amount of employment per each sector with an element Q_j (Table 2.1). In order to measure the amount of employment required for the production of one unit of total output, intensity coefficient matrix **q** is then calculated with elements q_j, where $q_j = Q_j/x_j$, Q_j hold a value of the employment satellite account into the production of sector j, and x_j is the total output of sector j.

The direct requirement matrix **A** of Table 2.1 can be expressed as

$$\mathbf{A} = (a_{ij}) = \begin{pmatrix} a_{11} & a_{12} & a_{13} \\ a_{21} & a_{22} & a_{23} \\ a_{31} & a_{32} & a_{33} \end{pmatrix} = \begin{pmatrix} T_{11}/x_1 & T_{12}/x_2 & T_{13}/x_3 \\ T_{21}/x_1 & T_{22}/x_2 & T_{23}/x_3 \\ T_{31}/x_1 & T_{32}/x_2 & T_{33}/x_3 \end{pmatrix} \tag{2.1}$$

The gross output of sector i in Table 2.1 can be written as follow:

$$x_i = \sum_{j=1}^{3} T_{ij} + y_i \quad \text{from Eq. 2.1} \quad T_{ij} = a_{ij}x_j \tag{2.2}$$

Table 2.1 Input-output framework coupled with an employment satellite account

		Intermediate demand			Final demand	
		Agriculture	Manufacturing	Services	Gross national expenditure	Exports
Intermediate inputs	Agriculture	T Intermediate matrix			***y*** Final demand	
	Manufacturing					
	Services					
Value added	Gross domestic products	***v*** Value added to production				
	Imports					
Satellite account	Employment (FTE)	Agriculture	Manufacturing	Services		

then

$x_i = \sum_{j=1}^{3} a_{ij}x_j + y_i$ if we require to consider the entire economy, each parameter can be expressed as:

$$\boldsymbol{y} = \begin{pmatrix} y_1 \\ y_2 \\ y_3 \end{pmatrix}; \quad \mathbf{X} = \begin{pmatrix} x_1 \\ x_2 \\ x_3 \end{pmatrix};$$

Using Eqs. 2.1 and 2.2 and the fundamentals of matrix multiplication, we can find that the matrix **T** is equal to

$$\mathbf{T} = (\mathrm{T}_{ij}) = (a_{ij}x_j) = \begin{pmatrix} a_{11}x_1 + a_{12}x_2 + a_{13}x_3 \\ a_{21}x_1 + a_{22}x_2 + a_{23}x_3 \\ a_{31}x_1 + a_{32}x_2 + a_{33}x_3 \end{pmatrix} = \begin{pmatrix} a_{11} & a_{12} & a_{13} \\ a_{21} & a_{22} & a_{23} \\ a_{31} & a_{32} & a_{33} \end{pmatrix} \begin{pmatrix} x_1 \\ x_2 \\ x_3 \end{pmatrix};$$

Thus,

$$\begin{pmatrix} x_1 \\ x_2 \\ x_3 \end{pmatrix} = \begin{pmatrix} a_{11} & a_{12} & a_{13} \\ a_{21} & a_{22} & a_{23} \\ a_{31} & a_{32} & a_{33} \end{pmatrix} \begin{pmatrix} x_1 \\ x_2 \\ x_3 \end{pmatrix} + \begin{pmatrix} y_1 \\ y_2 \\ y_3 \end{pmatrix}$$ and also can be written as

$$\mathbf{X} = \mathbf{AX} + \boldsymbol{y} \text{ and } \boldsymbol{y} = \mathbf{X} - \mathbf{AX} \Rightarrow \boldsymbol{y} = \mathbf{IX} - \mathbf{AX} \Rightarrow \boldsymbol{y} = (\mathbf{I} - \mathbf{A})\mathbf{X} \tag{2.3}$$

Multiplying both sides of Eq. 2.3[2] by$(\mathbf{I} - \mathbf{A})$ then

$$\mathbf{X} = (\mathbf{I} - \mathbf{A})^{-1}\boldsymbol{y} \tag{2.4}$$

While **A** has the direct requirement of the production, the factor $(\mathbf{I}-\mathbf{A})^{-11}$, or what is well-known the Leontief inverse **L** $[\mathbf{L} = (\mathbf{I} - \mathbf{A})^{-11}]$, has the direct and indirect links between sectors. Thus, the total output **X** in Eq. 2.4 can be represent as

$$\mathbf{X} = \mathbf{L}\boldsymbol{y} \tag{2.5}$$

By multiplying both sides of Eq. 2.5 by q, then $\mathbf{qX} = \mathbf{qL}\boldsymbol{y}$. Setting $m = \mathrm{qL}$, where m is a multiplier that show the amount of employment that is required directly (within a country) and indirectly (throughout a long supply chain) to satisfy the consumption of a population.

---

[2]**I** is an identity matrix. In matrix algebra, it has no effect on the equation and hence it has the same power as multiplying by one in an ordinary algebra.

References

Alsamawi, A., Murray, J., & Lenzen, M. (2014). The employment footprints of nations uncovering master-servant relationships. *Journal of Industrial Ecology, 18*(1), 59–70.

Alsamawi, A., Murray, J., & Lenzen, M. (2016). The working conditions footprints of nations. *Journal of Cleaner Production* (submitted).

Gomez-Paredes, J., Alsamawi, A., Yamasue, E., Okumura, H., Ishihara, K., Geschke, A., et al. (2016). Consuming childhoods: An assessment of child labor's role in Indian production and global consumption. *Journal of Industrial Ecology, 20*(3), 611–622.

Haberl, H., Erb, K.-H., Krausmann, F., Gaube, V., Bondeau, A., Plutzar, C., et al. (2007). Quantifying and mapping the human appropriation of net primary production in earth's terrestrial ecosystems. *Proceedings of the National Academy of Sciences of the United States of America, 104*(31), 12942–12947.

Hertwich, E. G., & Peters, G. P. (2009). Carbon footprint of nations: A global, trade-linked analysis. *Environmental Science and Technology, 43*(16), 6414–6420.

Leontief, W. (1936). Quantitative input and output relations in the economic system of the United States. *The Review of Economic and Statistics, 18,* 105–125.

Simas, M. S., Golsteijn, L., Huijbregts, M. A. J., Wood, R., & Hertwich, E. G. (2014). The "bad labor" footprint: Quantifying the social impacts of globalization. *Sustain, 6,* 7514–7540. doi:10.3390/su6117514

Wiedmann, T., Schandl, H., Lenzen, M., Moran, D., Suh, S., West, J., et al. (2013). The material footprint of nations—reassessing resource productivity. *Proceedings of the National Academy of Sciences of the United States of America, 110*(36), 1–6.

Chapter 3
Introduction to the Social Footprints of Global Trade

Accounting for social impacts within a supply chain can be a complex process. As trade barriers are reduced and global trade increases supply chains have become more complex and more opaque because of the number of parties that can be involved. Organisations need to understand not only where supply chains start and finish, but also the social (and environmental) impacts that occur along the way. The business drivers for understanding production and supply chains include reporting requirements, legal requirements, chain of custody certification, risk management, business to business and consumer pressure, logistics and efficiency and ethics. Businesses are finding that supply chain impacts are material to their operations and hence should be included in management indicators and organisational reporting.

Increasingly, corporate responsibility and legislation will require accounting for sustainability in production and supply chains, particularly those extending into high social (and environmental) risk areas. Although there has been a greater focus in literature on environmental impacts of supply chains, it is the social impacts that tend to create a negative image that companies can find difficult to shake off. One of the first major publicity scandals relating to social impacts in a global supply chain was that of the sporting brand Nike. The use of child labour in its production supply chain for the sewing of footballs is well known. Many years after this case was first publicised, its legacy remains in the public perception even though Nike has since put in place processes to ensure that labour within their supply chain is in line with if not better than international labour standards (Birch 2012).

Even for sustainability reporting businesses now need to focus on the impacts within their supply chains. The G4 Reporting Guidelines released in 2013 by the Global Reporting Initiative[1] have an increased emphasis on materiality and supply chain management compared with previous guidelines, with supply chain disclosures now required as standard. Relatively few tools enable the analysis of global production and supply chains, and even fewer offers the ability to analyse multiple

[1]https://www.globalreporting.org/reporting/g4.

A. Alsamawi et al., *The Social Footprints of Global Trade*,
Environmental Footprints and Eco-design of Products and Processes,
DOI 10.1007/978-981-10-4137-2_3

supply chains at once. The purpose of this book is to demonstrate a methodology to help determine the scale and existence of social impacts within global supply chains. Where social data exist sourced from research into production (for example, the number of hours worked by children to stitch footballs) how can it be used to quantify the responsibility for impact in consumption? We can get the upstream data, usually through means such as social audits, surveys or questionnaires. What we do not understand is how much of that social impacts we consume are responsible for consuming, once the product has gone through numerous supply chain paths to different countries. This book showcases a technique to connect social impacts occurring upstream with the countries of consumption downstream. It provides examples of how social impacts can be measured and tracked through supply chains, using Multi-Regional Input-Output Analysis (MRIOA). This introduction will examine current methods for accounting for social impacts in supply chains and explain in greater detail what MRIOA is.

3.1 Measuring Social Impacts in Supply Chains

The social responsibility of business has many interpretations. These range from Milton Friedman's assertion that business has no social responsibility, only responsibility to increase profitability within the rules of the game (Friedman 1970) to John Elkington redefining business accounting through a triple bottom line approach (Elkington 1998) and many theories in between and beyond. Gray argues that there is no evidence to support liberal economics as described by Friedman, and that the economic system of accounting and reporting on which it is based has fundamental flaws, as evidenced, for example, by the increasing inequality gap between rich and poor and the decline of the state of the environment (Gray 2006; 796). The social responsibility of business is often referred to as corporate social responsibility. Corporate social responsibility (CSR) was simply defined by McWilliams and Siegal (2001: 117) as "*...actions which appear to further some social good, beyond the interests of the firm and that which is required by law*". Contributing to the theory of CSR was Freeman's work on the management of stakeholder expectations for strategic outcomes and subsequent development of the ideas around stakeholder identification (Freeman 1984; Mitchell et al. 1997). If we assume that businesses do have a responsibility for the social impacts they generate (through supply or employment), then robust assessment and accounting methodologies are required.

There is increasing importance being placed on accounting for sustainability impacts in supply chains (Wolf 2011; Erol et al. 2011; Quadra et al. 2009; Corbiere-Nicollier et al. 2011). The benefits of managing supply chain impacts have been well noted for competitive advantage, improved efficiency and operations (Lee 2002; Kim 2006, 2009; Klibi et al. 2010; Lainez et al. 2009; Mefford 2009; Mosovsky et al. 2000; Nyaga et al. 2010; Prado-Prado 2009). To date, most of the research on this subject has been into 'greening' supply chains or improving environmental outcomes. Social accounting is most frequently addressed as part of social and environmental accounting (SEA) or sustainability accounting, rather than addressing social accounting on its own. Shaw et al. (2010) found over 2000 references on the ISI Web of Science database when searching for 'environmental supply chain performance'. The body of work on managing the social sustainability of supply chains is smaller, but increasing. It often focuses on how businesses manage their supply chain accountability, accounting and reporting, linking CSR activities with supply chain management (Spence and Bourlakis 2009; Kortelainen 2008; Awaysheh and Klassen 2010).

In their assessment of sustainability accounting and reporting, Burritt and Schaltegger (2010) divide the literature into two main bodies: the critical theorists who propose that sustainability accounting does nothing to support planetary sustainability, and that reporting can in fact deflect away from or even obfuscate the true destruction caused by business in the pursuit of economic growth; and those who support the development of sustainability accounting if it provides internal or external stakeholders with 'useful and high quality' information (p. 833) for managerial focussed accounting. However, data provided for sustainability accounting is not always useful or high quality. There is a lack of quantification of social accounting, in particular. In many instances, sustainability reporting is used purely to drive branding message rather than demonstrate real commitment to improved environmental or social outcomes. Hopwood illustrates 'green advertising' using the example of oil companies and the rhetoric between their actions and words (Hopwood 2009: 438–439).

A gap in this body of work is that few SEA methodologies rely on economic accounting tools to account for social and environmental impacts, despite purporting to be accounting. The time has come to fill this gap, and this book will demonstrate a way in which economic tools can be used to account for social impacts. In their review of the past 20 years of development of SEA literature, Gray and Laughlin (2012) do not discuss the evolution of input-output analysis (IOA) as a tool for social or environmental accounting. IOA is used in some forms of social accounting (such as social accounting matrices) but the use of environmentally and socially extended MRIOA is relatively new. If financial reporting and SEA "derive from fundamentally different views of the world" (Gray 2006: 794) then surely the use of socially and environmentally extended MRIOA can go some way towards bringing these worlds together and bridging the gap. MRIOA is based on economic accounts and trade flows, usually in US dollars. The ability to associate physical accounts (such as carbon emissions or deaths) as inputs to economic trade is part of the ingenuity of MRIOA. In his ground-breaking work developing the input-output framework (that was eventually recognised with a Nobel Prize in 1973), Wassily Leontief realised that you could use input-output analysis to model how much pollution input was required to produce a certain output (a car, for example). In this way we explore here how social impacts are also an input to production.

Boundary definition and data availability are difficulties that arise with many methods of social or environmental analysis. For example some of the challenges for social life cycle assessment are data availability and the definition of the assessment scope (Lehmann et al. 2013). Data requirements and boundaries are an important consideration for accuracy in SEA (Burritt and Schaltegger 2010; Gray and Laughlin 2012; Aras and Crowther 2009). A single physical supply chain audit of a global supply chain will be limited by time, availability, cost and access. However, MRIOA can take into account over five million supply chains in one calculation from upstream production to downstream consumption. Boundaries do need to be drawn, but not as tightly as required with most other methodologies. This power of MRIOA has only become available since around 2008, because of improved computational powers used in the construction and data population of global MRIO databases (Kanemoto and Murray 2013).

Hopwood (2009) argues that calculation and new forms of accounting should be a feature of achieving a greater harmony between human endeavours and planetary sustainability. Hopwood uses the carbon emissions market as an example of a link between accounting and the environment. MRIO models are regularly used to model both consumer and producer carbon emissions because the methodology allows for calculation of a complete carbon footprint (Wiedmann 2009). In this book we demonstrate the use of this methodology applied to social accounting, using the social impacts of work instead of greenhouse gas emissions to develop a social footprint. In doing so we show that socially-extended MRIO accounts provide the accounting field with a tool that has an accounting/economics foundation that can be combined with qualitative analysis including case study analysis and qualitative textual analysis to understanding the impacts of global trade and its supply chains.

References

Aras, G., & Crowther, D. (2009). Corporate sustainability reporting: A study in disingenuity? *Journal of Business Ethics, 87*(1), 279–288.

Awaysheh, A., & Klassen, R. (2010). The impact of supply chain structure on the use of supplier socially responsible practices. *International Journal of Operations & Production Management, 30*(12), 1246–1268.

Birch, S. (2012). How activism forced Nike to change its ethical game. *The Guardian*, p.http://www.theguardian.com/environment/green-living-blog/2012/jul/06/activism-nike

Burritt, R., & Schaltegger, S. (2010). Sustainability accounting and reporting: Fad or trend? *Accounting Auditing & Accountability Journal, 23*(7), 829–846.

Corbiere-Nicollier, T., Blanc, I., & Erkman, S. (2011). Towards a global criteria based framework for the sustainability assessment of bioethanol supply chains application to the Swiss dilemma: Is local produced bioethanol more sustainable than bioethanol imported from Brazil? *Ecological Indicators, 11*(5), 1447–1458.

Elkington, J. (1998). *Cannibals with forks: The triple bottom line of 21st century business.* Gabriola Island: New Society.

Erol, I., Sencer, S., & Sari, R. (2011). A new fuzzy multi-criteria framework for measuring sustainability performance of a supply chain. *Ecological Economics, 70*(6), 1088–1100.

Freeman, R. E. (1984). *Strategic management: A stakeholder perspective*. NJ, Prentice Hall: Englewood Cliffs.

Friedman, M. (1970). The social responsibility of business is to increase its profits. *The New York Times Magazine,* (September 13).

Gray, R. (2006). Social, environmental and sustainability reporting and organisational value creation? *Accounting, Auditing & Accountability Journal, 19*(6), 793–819.

Gray, R., & Laughlin, R. (2012). It was 20 years ago today. *Accounting, Auditing & Accountability Journal, 25*(2), 228–255.

Hopwood, A. G. (2009). Accounting and the environment. *Accounting, Organizations and Society, 34*(3–4), 433–439.

Kanemoto, K., & Murray, J. (2013). What is MRIO: Benefits and Limitations. In J. Murray & M. Lenzen (Eds.), *The sustainability practitioner's guide to multi-regional input-output analysis.* Illinois, USA: Common Ground.

Kim, S. W. (2006). Effects of supply chain management practices, integration and competition capability on performance. *Supply Chain Management-An International Journal, 11*(3), 241–248.

Kim, S. W. (2009). An investigation on the direct and indirect effect of supply chain integration on firm performance. *International Journal of Production Economics, 119*(2), 328–346.

Klibi, W., Martel, A., & Guitouni, A. (2010). The design of robust value-creating supply chain networks: A critical review. *European Journal of Operational Research, 203*(2), 283–293.

Kortelainen, K. (2008). Global supply chains and social requirements: Case studies of labour condition auditing in the People's Republic of China. *Business Strategy and the Environment, 17,* 431–443.

Lainez, J. M., Reklaitis, G. V., & Puigjaner, L. (2009). Managing financial risk in the coordination of supply chain and product development decisions. *19th European Symposium on Computer Aided Process Engineering,* Vol. 26, pp. 1027–1032.

Lee, H. (2002). Aligning supply chain strategies with product uncertainties. *California Management Review, 44*(3), 105–119.

Lehmann, A., Zschieschang, E., Traverso, M., Finkbeiner, M., & Schebek, L. (2013). Social aspects for sustainability assessment of technologies—Challenges for social life cycle assessment (SLCA). *The International Journal of Life Cycle Assessment, 18*(8), 1581–1592.

McWilliams, A., & Seigal, D. (2001). Corporate social responsibility: A theory of the firm perspective. *Academy of Management Review, 26*(1), 117–127.

Mefford, R. N. (2009). Offshoring, lean production, and a sustainable global supply chain. *Icoscm 2009—Proceedings of the 3rd International Conference on Operations and Supply Chain Management,* Vol. 3, pp. 268–275.

Mitchell, R. K., Agle, B. R., & Wood, D. J. (1997). Toward a theory of stakeholder identification and salience: Defining the principle of who and what really counts. *The Academy of Management Review, 22*(4), 853–886.

Mosovsky, J., Dickinson, D., & Morabito, J. (2000). Creating competitive advantage through resource productivity, eco-efficiency, and sustainability in the supply chain. *Proceedings of the 2000 IEEE International Symposium on Electronics and the Environment*, pp. 230–237.

Nyaga, G. N., Whipple, J. M., & Lynch, D. F. (2010). Examining supply chain relationships: Do buyer and supplier perspectives on collaborative relationships differ? *Journal of Operations Management, 28*(2), 101–114.

Prado-Prado, J. C. (2009). Continuous improvement in the supply chain. *Total Quality Management & Business Excellence, 20*(3), 301–309.

Quadra, P. R. B., Rodriguez, J. M., Terol, M. C. R., & Cruz, D. E. (2009). A three level multi-period multi-location and multi-crop sustainable supply chain model. *2009 IEEE International Conference on Industrial Engineering and Engineering Management,* Vol. 1–4, pp. 1414–1418.

Shaw, S., Grant, D. B., & Mangan, J. (2010). Developing environmental supply chain performance measures. *Benchmarking: An International Journal, 17*(3), 320–339.

Spence, L., & Bourlakis, M. (2009). The evolution from corporate social responsibility to supply chain responsibility: The case of Waitrose. *Supply Chain Management, 14*(4), 291–302.

Wiedmann, T. (2009). A review of recent multi-region input–output models used for consumption-based emission and resource accounting. *Ecological Economics, 69*(2), 211–222.

Wolf, J. (2011). Sustainable supply chain management integration: A qualitative analysis of the german manufacturing industry. *Journal of Business Ethics, 102*(2), 221–235.

Chapter 4
Review of Social Accounting Methodologies

4.1 Supplier Social Audits

Supplier social audits are generally conducted on behalf of the purchaser of a product or service. They involve sending a representative (either directly employed or a third party) to a place of production within the supply chain. These audits often take place in factories (Kortelainen 2008) but can also travel down the supply chain to primary producers, particularly in the food or textile industries (Miller 2011; Spence and Bourlakis 2009). As Hannah Jones, Head of Global Corporate Social Responsibility at Nike is quoted as saying of social audits "*All monitoring does is reveal the issues. It doesn't solve them. The reasons for excessive overtime, for example, are horribly complicated. You have to do system analysis. You have to do 'root cause' analysis.*" (Foley 2012). However, revealing the issues and what is happening on the ground is a significant advantage of this method over all other methods discussed.

4.2 Third Party Certification and Verification

Another methodology for pursuing social accountability in supply chain management by business is using a third party certification or chain of custody certification to established standards. For example, Fairtrade is a third party certification process that focuses on guaranteed prices and codes of conduct as a system for sustainable supply chain management (Welford et al. 2003; Vermeulen and Seuring 2009). Compliance with the key International Labour Organisation (ILO) conventions is commonly found in standards with a social impact focus such as Fairtrade, Ethical Trading Initiative, United Nations Global Compact and the Forestry Stewardship Council standard. Chain of custody certification can be costly and time consuming (Suryani et al. 2011) with varying results because of the different certification

A. Alsamawi et al., *The Social Footprints of Global Trade*,
Environmental Footprints and Eco-design of Products and Processes,
DOI 10.1007/978-981-10-4137-2_4

systems available and their requirements. The benefit of chain of custody certification or third party verification is that it can enable a buyer (or consumer) to know that each link of a supply chain has been verified to a certain known standard, or know that the production of a good has been audited to a known standard, without actually physically auditing the supply chain (as outlined in the previous section). The standards tend to be developed by the industry as well as social and environmental non-governmental organisations (NGOs) and as a consequence have requirements for high environmental and social outcomes.

4.3 Purchasing and Procurement Practices

Purchasing and procurement practices are other approaches used for accountability in supply chain management to gather supply chain information and control supply chain practices. Roberts (2003) highlights that there has been an increasing trend to outsource business processes and activities to suppliers and subcontractors, placing responsibility for corporate governance aspects such as environmental and social impact risk management within procurement functions. Standards, such as BS8903: 2010—Principles and Framework for Procuring Sustainably, non-governmental organisation (NGO) and government advice are increasingly providing guidance to organisations on how to take social issues into account in the procurement process. For example, the United Nations Global Compact, a code of conduct containing social standards aimed at improving CSR, will have an influence on business supply chain management and accountability (Leisinger 2007; Garsten and Jacobsson 2011; Marinilka and Cao 2011; Janney et al. 2009; Rasche 2009). However studies have also shown that using CSR policies in supply chain programs can have a negative effect on businesses, particularly on small to medium enterprises, due to issues such as increased cost and time burdens (Ciliberti et al. 2011; Villena et al. 2011; Baden et al. 2011). The benefits of using purchasing and procurement policies are that they are lower cost and easier to implement because they generally don't involve site visits or third party verification (although both of these can still take place). They can also be applied easily across a number of supply chains.

4.4 Multi-regional Input–Output Analysis

MRIOA is the use of multiple input-output tables, arranged in a matrix database, to analyse economic trades and flows across multiple countries or regions. Individual input-output tables demonstrate, on a monetary basis, the inputs to industry sectors required to produce goods or services and the outputs of each industry sector, and the interrelationships for one country or region. Thus an input-output table shows the flows of money through different sectors of the economy to final consumption

(Murray and Wood 2010). Most countries produce their own input-output tables at regular intervals, with data collected in accordance with the System of National Accounts (United Nations Statistics Division 2009). MRIO databases have evolved as computing power increases and are usually developed by research centres, with less than ten databases currently in development and use globally (Murray and Lenzen 2013).

Several different methods of analysis have been used in this book using of MRIOA footprinting. In this chapter we used sector disaggregation to provide supply chain analysis as well as associating a social satellite account. A satellite account is an account containing relevant data that is linked to but separate from the main IO table/s (United Nations Statistics Division 2008), generally linking physical (or non-monetary) accounts such as water or carbon emissions to economic (monetary) accounts. Disaggregating a sector using detailed data gathered through research makes it possible to trace a very specific supply chain through an economy using input-output analysis (Lenzen 2011; Liu et al. 2012). The methodology used in this chapter follows the money flows from one industry sector to the other through to final consumption using structural path analysis, as used by Wood and Lenzen (2009), Suh and Heijungs (2007) and Lenzen (2006) to analyse supply chains.

There are many uses for MRIOA. Examples include assessing environmental impacts such as the energy, emissions or embodied carbon intensity of sectors of the economy or industry sectors (Treloar 1997; Virtanen et al. 2011; Zhou 2010; Liu et al. 2012), and consumption based accounting, particularly in reference to accounting for Greenhouse Gas Emissions (Wiedmann 2009). MRIOA is also well established for assessing policy decisions and alternative scenarios, such as assessing the costs and benefits of alternative forms of energy (Wiedmann et al. 2011), producer/consumer responsibility (Lenzen et al. 2007) and consumption and production (Wiedmann and Barrett 2011). However, the application of MRIOA to social supply chain analysis has to date been limited. The UNEP/SETAC guidelines on social LCA and the associated methodological sheets (Benoit-Norris et al. 2011) make a significant contribution to understanding the detailed social impacts of products, but not necessarily whole supply chains.

The use of satellite accounts for special purposes in input-output analysis, such as measuring health or environmental impacts, is outlined in the System of National Accounts (United Nations Statistics Division 2008; Chap. 29). Different approaches to assessing social impacts using combined LCA and MRIOA have been used, such as by Rugani et al. (2012). Studies linking MRIOA and social impacts are emerging particularly in relation to inequality between global trading partners (Alsamawi et al. 2014; Prell et al. 2014) and assessing product supply chains using the GTAP MRIO database and associated Social Hotspot Database (Benoit-Norris et al. 2012). However the combination of using sector disaggregation, social satellite accounts and structural path analysis has not been common to MRIOA for revealing detailed social supply chain impacts. This book demonstrates the value of MRIOA for better understanding social impacts in supply chains and for evaluating the social impacts of global trade.

There are many benefits of using MRIOA for supply chain analysis. The ability to trace millions of supply chains throughout the world has come about through the painstaking creation of detailed MRIO databases based on national economic accounts, and satellite accounts developed through other data sources such those collected by the World Bank, World Health Organisation, International Labour Organisation. To trace these supply chains through economic data without the use of MRIOA would be almost impossible, and certainly very time consuming. The scope provided by MRIOA is very large. Using economic data can provide rigor to the process of tracing social impacts in supply chains.

Another benefit of using a MRIO database is the ability to conduct an analysis with only partial data. The availability of a small amount of detailed data will provide a better result than its absence. This represents the combination of life cycle data with MRIOA. The inclusion of detailed data upstream in the supply chain enables this information to be allocated proportionally along the supply chain and accounted for downstream at the point of consumption.

The cost of research using an MRIO model (where available) is significantly less than that required for a physical supply chain audit, and may reveal enough information to allow business or government to focus their efforts. The applicability of MRIOA to a range of sustainability issues in supply chain accounting is significant—the approach can be used to model the supply chain of products of concern, such as palm oil, or to model supply chains where production methods are of concern, such as child labour. The process is already in use for the quantification of environmental impacts and lends itself well to further application in investigation of social supply chain issues particularly at an industry level.

There are, however, limitations to use of MRIOA for supply chain analysis. Limitations with input-output analysis are well understood. For example, if a business understands its size in proportion to the industry sector, it can estimate responsibility proportionally for the chosen indicator. However, if a business does not behave in an average manner (e.g. it uses only recycled material inputs to production) the results will need to be adjusted accordingly. IOA is based on macroeconomic data, collected in accordance with the SNA (United Nations Statistics Division 2009). A nation's accounts depend on that nation being able to accurately collate data relating to labour, employment, income, taxes, trade and even population. However, where a country has a significant informal (or black market) economy or a population employed in non-paid work (e.g. domestic labour, subsistence farming), the SNA is unable to account for that economic activity (Waring 1988). Data inaccuracies will also affect this (e.g. under-reporting of workplace accidents). This limitation is in part addressed by the use of superior data, providing more detailed input on social impacts than are otherwise unaccounted for in national accounts.

It is important to note that most IO tables are in monetary terms (usually US$ for MRIO databases). The results thus can suffer from bias from price effects (e.g. using monetary analysis the environmental impacts of a $50 special fare for a flight will be reported as 1/10th that of the same flight at its normal $500 fare). Flows into and out of some financial sectors, for example, may be reported as highly environmentally or socially intensive even though the actual physical flows into/out of these sectors are in fact small.

Further discussions on IO analysis limitations and strengths in general such as sector averages, uncertainties of data, scope and double counting can be found in texts such as by Murray and Wood (2010). Further details regarding the data quality of the Eora MRIO tables, including data optimisation, standard deviation settings and the confidence of UN Main Aggregates and Official Country Data and UN Comtrade data can be found in Lenzen et al. (2012).

As with most data analysis, MRIOA is retrospective. Information provided by national statistical organisations may be two to five years behind the current year. As with any computational method, researchers must understand the information being analysed and fed into the system for analysis, as well as have the ability to interpret the results. However, MRIOA has been used successfully over defined time periods to help identify trends and predict future activity e.g. (Barrett et al. 2013).

Unlike physical audits, verified certification or even working with supply chain partners through procurement processes, MRIOA does not reveal the actual social situation on the ground. Without actual data collected from sites of interest, whether by the researchers themselves or other organisations (most often intergovernmental or NGO organisations), assumptions must be made, which reduces the certainty of the results. Where analysis is used with superior data certainty, for example as provided directly by an industry sector or by an interested NGO, the results of analysis will have a higher degree of certainty. This point brings together the importance of collaboration for MRIOA—collaboration in the methodology development, collaboration in getting detailed data, collaboration in sharing results and collaboration for using those results for human development.

References

Alsamawi, A., Murray, J., Lenzen, M., Moran, D., & Kanemoto, K. (2014). The inequality footprints of nations: A novel approach to quantitative accounting of income inequality. *PLOS ONE, 9*(10), e110881. doi:10.1371/journal.pone.0110881

Baden, D., Harwood, I. A., & Woodward, D. G. (2011). The effects of procurement policies on 'downstream' corporate social responsibility activity: Content-analytic insights into the views and actions of SME owner-managers. *International Small Business Journal, 29*(3), 259–277.

Barrett, J., Peters, G., Wiedmann, T., Scott, K., Lenzen, M., Roelich, K., et al. (2013). Consumption-based GHG emission accounting: A UK case study. *Climate Policy, 13*(4), 451–470.

Benoit-Norris, C., Cavan, D. A., & Norris, G. (2012). Identifying social impacts in product supply chains: Overview and application of the social hotspot database. *Sustainability, 4,* 1946–1965.

Benoit-Norris, C., Vickery-Niederman, G., Valdivia, S., Franze, J., Traverso, M., Ciroth, A., et al. (2011). Introducing the UNEP/SETAC methodological sheets for subcategories of social LCA. *International Journal of Life Cycle Assessment, 16*(7), 682–690.

Bullard, C. W., Penner, P. S., & Pilati, D. A. (1978). Net energy analysis—Handbook for combining process and input-output analysis. *Resources and Energy, 1,* 267–313.

Ciliberti, F., de Haan, J., de Groot, G., & Pontrandolfo, P. (2011). CSR codes and the principal-agent problem in supply chains: Four case studies. *Journal of Cleaner Production, 19* (8), 885–894.

Foley, S. (2012). Apple must learn from Nike and get tough on causes of supply chain abuse. *The Independent.*

Garsten, C., & Jacobsson, K. (2011). Transparency and legibility in international institutions: The UN Global Compact and post-political global ethics. *Social Anthropology, 19*(4), 378–393.

Janney, J. J., Dess, G., & Forlani, V. (2009). Glass houses? Market reactions to firms joining the UN global compact. *Journal of Business Ethics, 90*(3), 407–423.

Kortelainen, K. (2008). Global supply chains and social requirements: Case studies of labour condition auditing in the People's Republic of China. *Business Strategy and the Environment, 17,* 431–443.

Leisinger, K. M. (2007). Capitalism with a human face: The UN global compact. *The Journal of Corporate Citizenship, 28,* 113–132.

Lenzen, M. (2011). Aggregation versus disaggregation in input-output analysis of the environment. *Economic Systems Research, 23*(1), 73–89.

Lenzen, M., Geschke, A., Kanemoto, K & Moran, D. (2012). Mapping the structure of the World Economy. *Environmental Science & Technology, 46*(15), 8374–8381.

Lenzen, M., Murray, J., Sack, F., & Wiedmann, T. (2007). Shared producer and consumer responsibility—Theory and practice. *Ecological Economics, 61*(1), 27–42.

Lenzen, M. (2006). Structural path analysis of ecosystem networks. *Ecological Modelling, 200*(3–4), 334–342.

Liu, C.-H., Lenzen, M., & Murray, J. (2012). A disaggregated emissions inventory for Taiwan with uses in hybrid input-output life cycle analysis (IO-LCA). *Natural Resources Forum, 36* (2), 123–141.

Marinilka B. K., & Cao, Z. (2011). Does voluntary corporate citizenship pay? An examination of the UN Global Compact. *International Journal of Accounting and Information Management, 19*(3), 288–303.

Miller, Z. (2011). Euromonitor international ethical fashion series: Marks & Spencer boosts profitability through sustainable practices. *Euromonitor International* (Online).

Murray, J., & Lenzen, M. (Eds.). (2013). *The sustainability practicioner's guide to multi-regional input-output analysis.* Champaign, Illinois, USA: Common Ground.

Murray, J., & Wood, R. (Eds.). (2010). *The sustainability practicioner's guide to input-output analysis.* New York: Common Ground Publishing LLC.

Prell, C., Feng, K., Sun, L., Geores, M., & Hubacek, K. (2014). The economic gains and environmental losses of US consumption: A world-systems and input-output approach. *Social Forces.*

Rasche, A. (2009). Toward a model to compare and analyze accountability standards—The case of the UN Global Compact. *Corporate Social Responsibility and Environmental Management, 16* (4), 192–205.

Roberts, S. (2003). Supply chain specific? Understanding the patchy success of ethical sourcing initiatives. *Journal of Business Ethics, 44*(2–3), 159–170.

Rugani, B., Panasiuk, D., & Benetto, E. (2012). An input–output based framework to evaluate human labour in life cycle assessment. The International Journal of Life Cycle Assessment, 17 (6), 795–812.

Spence, L., & Bourlakis, M. (2009). The evolution from corporate social responsibility to supply chain responsibility: The case of Waitrose. *Supply Chain Management, 14*(4), 291–302.

Suh, S. (2004). Functions, commodities and environmental impacts in an ecological-economic model. *Ecological Economics, 48,* 451–467.

Suh, S., & Heijungs, R. (2007). Power series expansion and structural analysis for Life Cycle Assessment. *International Journal of Life Cycle Assessment, 12*(6), 381–390.

Suh, S., & Nakamura, S. (2007). Five years in the area of input-output and hybrid LCA. *International Journal of Life Cycle Assessment, 12*(6), 351–352.

Suryani, A. G. N., Shahwahid, H. O. M., Fauzi, P. A., Alias, R., & Vlosky, R. P. (2011). Assessment of chain-of-custody certification costs for sawnwood manufacturers in Peninsular Malaysia. *Journal of Tropical Forest Science, 23*(2), 159–165.

Treloar, G. (1997). Extracting embodied energy paths from input-output tables: Towards an input-output-based hybrid energy analysis method. *Economic Systems Research, 9*(4), 375–391.

United Nations Statistics Division. (2009). *System of national accounts 2008.*

United Nations Statistics Division. (2008). *The system of national accounts (SNA).*

Vermeulen, W. J. V., & Seuring, S. (2009). Sustainability through the market—The impacts of sustainable supply chain management: Introduction. *Sustainable Development, 17*(5), 269–273.

Villena, V. H., Revilla, E., & Choi, T. Y. (2011). The dark side of buyer-supplier relationships: A social capital perspective. *Journal of Operations Management, 29*(6), 561–576.

Virtanen, Y., Kurppa, S., Saarinen, M., Katajajuuri, J. M., Usva, K., Maenpaa, I., et al. (2011). Carbon footprint of food-approaches from national input-output statistics and a LCA of a food portion. *Journal of Cleaner Production, 19*(16), 1849–1856.

Waring, M. (1988). *If women counted: A new feminist economics.* San Francisco: Harper Collins Publishers.

Welford, R., Meaton, J., & Young, W. (2003). Fair trade as a strategy for international competitiveness. *International Journal of Sustainable Development and World Ecology, 10*(1), 1–13.

Wiedmann, T., & Barrett, J. (2011). A greenhouse gas footprint analysis of UK Central Government, 1990–2008. *Environmental Science & Policy.*

Wiedmann, T. O., Suh, S., Feng, K., Lenzen, M., Acquaye, A., Scott, K., et al. (2011). Application of hybrid life cycle approaches to emerging energy technologies—The case of wind power in the UK. *Environmental Science and Technology, 45*(13), 5900–5907.

Wiedmann, T. (2009). A review of recent multi-region input–output models used for consumption-based emission and resource accounting. *Ecological Economics, 69*(2), 211–222.

Wood, R., & Lenzen, M. (2009). Structural path decomposition. *Energy Economics, 31*(3), 335–341.

Zhou, X. (2010). Environmentally extended multi-region input-output model: Sharing responsibility across the globe. In J. Murray & R. Wood (Eds.), *The sustainability practitioners guide to input output analysis* (pp. 106–129). Common Ground Publishing.

Chapter 5
Review of Social Metrics and Social Footprinting

5.1 Introduction: Social Indicators and Metrics

Many indicators of social impact and social progress have been developed over the past 50 years, particularly as alternatives to economic indicators such as Gross Domestic Product (GDP) or measures of subjective wellbeing. For example, a 2013 study showed that although GDP has steadily increased since 1950, life satisfaction (as measured by the Genuine Progress Indicator) has not, and indeed it flattened out in the late 1970s and has not recovered (Kubiszewski et al. 2013). The indicators and metrics attempt to demonstrate, usually as an agglomeration of indicators, either societal progress by country or personal (subjective) wellbeing. Examples of societal wellbeing indices that compare social wellness and progress between countries include the Human Development Index, the OECD Better Life Index, the Well Being of Nations and the National Wellbeing Index. Indices that compare year-on-year country progress include the Australian Unity Well-being Index, the Canadian Index of Well-being and Gross National Happiness. Table S1 in the supplementary information to the paper by Costanza et al. (2014) provides a good summary of alternative national indicators of welfare and wellbeing.

Taking a more detailed look at these indices, one of the earliest composite well-being indicators is Bhutan's Gross National Happiness (GNH) indicator. Established in 1972, it measures progress against nine domains (psychological wellbeing, time use, community vitality, cultural diversity, ecological resilience, living standard, health, education, good governance). Employment is not one of the overarching domains, although it is covered within Time Use. The GNH indicator recognises that wellbeing requires a balance between paid work, unpaid work and leisure time, and thus respects a limit of eight hours of work per day, including unpaid work such as child-care and voluntary work. Those who work over eight

A. Alsamawi et al., *The Social Footprints of Global Trade*,
Environmental Footprints and Eco-design of Products and Processes,
DOI 10.1007/978-981-10-4137-2_5

hours a day are identified for the purposes of the GNH indicator as time deprived (Ura et al. 2012). In a similar way, the European Working Time Directive[1] aims to limit working time, to protect people's health and safety. EU directive also limits people to work less than 35 h a week, or just under eight hours per day. Many indicators and statistics are collected to monitor and demonstrate the social impacts of the workplace. Datasets from the International Labour Organisation[2] in particular provide data on the many dimensions of the social impact of work on people, including LABORSTA (database of labour statistics up until 2008), ILOSTAT (providing statistics on over 100 indicators) which exist to show when there are negative impacts within the workplace.

To develop a better understanding of social impacts in trade and development, an EU study analysed the social risk profile of EU-27 trade regarding trading partners in 2010. This research covered five thematic areas: Labour Rights and Decent Work; Health and Safety; Human Rights; Governance; and Community Infrastructure (Pelletier et al. 2013). The EU research was in part in recognition that the EU's external policies must respect the '*principles of democracy, the rule of law, the universality and indivisibility of human rights and fundamental freedoms, respect for human dignity, the principles of equality and solidarity, and respect for the principles of the United Nations Charter and international law*' (EC 2008[3]). The thematic areas identified in the EU study aligned with the thematic areas of the Social Hotspot Data base.

The Social Hotspot database[4] uses the primary social impact measurement tools available at an international level—the UNEP/SETAC Guidelines for Social Life Cycle Assessment (UNEP/SETAC 2009), the ISO 26000 Guidelines for Social Responsibility (ISO 2010), the Global Reporting Initiative (GRI) Guidelines and the Global Social Compliance Programme Reference tools (Benoit-Norris et al. 2012). Figure 5.1 presents a social hotspot index for five of the developing economies considered in the study—Brazil, China, India, Russian Federation, Madagascar, three of which (China, India and Madagascar) have been identified as net exporters of labour (Alsamawi et al. 2014, Table 2). Each circle represents the proportional contribution of labour rights and decent work for the food production industry (vegetables, fruits and nuts) in that country as calculated by social LCA. The map shows, for example, that labour rights and decent work is an issue for all of these countries, and health and safety is a significant concern in Brazil and India.

---

[1]Directive 2003/88/EC of the European Parliament and of the Council of 4 November 2003 concerning certain aspects of the organisation of working time http://eur-lex.europa.eu/LexUriServ/LexUriServ.do?uri=CELEX:32003L0088:EN:HTML.

[2]http://www.ilo.org/global/statistics-and-databases/lang–en/index.htm.

[3]EC 2008a. Consolidated Version of the Treaty on the Functioning of the European Union. Official Journal of the EU C115: 47-199.

[4]www.socialhotspot.org.

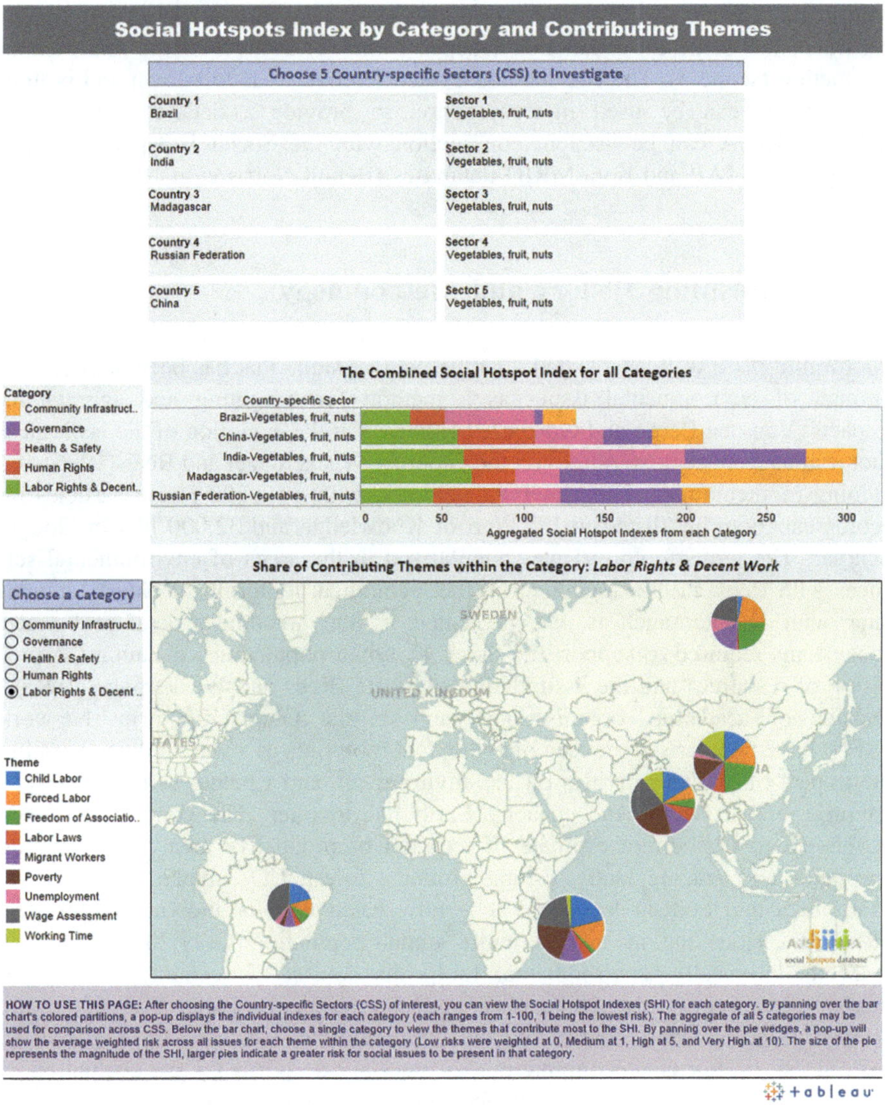

Fig. 5.1 Social hotspots for developing countries in agricultural sectors. *Source* Social Hotspot Database Index, accessed 01/08/16

Using the social LCA methodology, the contributing themes within each category are shown for the social impact within that country (domestic labour). This differs for the MRIOA footprinting methodology, where the footprint is calculated based on both the domestic labour from within that country, as well as the imported labour from upstream in the production chain to provide a social footprint. If we draw an analogy with carbon footprinting and the definitions drawn for scopes 1, 2 and 3 within the Greenhouse Gas Protocol,[5] the social footprint calculated in Chap. 6 considers the domestic production (scopes 1 and 2) as well as the value chain production (scope 3). The two methods, often referred to as top down and bottom up analysis, can be used in conjunction to provide a detailed assessment. Indeed MRIOA can be used in conjunction with the Social Hotspots database through the GTAP and Eora MRIO databases (Benoit-Norris et al. 2012).

5.2 Footprinting History and Methodology

Footprinting is a well-established quantitative technique that has been applied to a number of environmental issues as a method for weighting and aggregating impacts. Van den Bergh and Grazi (2014) looked at the evolution of the ecological footprint as a concept, starting with the paper by Wackernagel and Rees (1996) and gaining momentum over the intervening years, with over 500 journal articles on ecological footprints listed in ISI Web of Knowledge and 32,000 hits in Google Scholar. The term 'footprint' was popularized in the arena of environmental sciences with Rees' (1992) argument on the 'ecological footprint' of cities. Matured later with Wackernagel, it was presented as the amount of land and water ecosystems required to support the 'load' of urban populations, meaning the provision of resources and the assimilation of waste (Rees and Wackernagel 1996).[6] Outside of academia, organisations such as the Global Footprint Network (GFN) have popularised the idea of ecological footprints as a way of demonstrating the impact of human activities on the environment, and whether as individuals or countries we fit within the capacity Earth has to offer. The concept has been developed to the current estimate of required bioproductive area, encompassing croplands and grazing lands, fishing grounds, forest, CO_2 uptake area ('carbon footprint'), and built-in land; and it is now calculated for individual lifestyles, businesses, cities, nations and the entire human population (GFN 2014).

Methodologically, environmental footprints consist of accounts of physical flows, often being time demanding and data limited. Then, Bicknell et al. (1998) proposed their estimation via Input–Output Analysis (IOA); that is, Leontief's (1936) macroeconomic model precisely designed for the evaluation of *total* impacts of economic production/consumption. By then, IOA had already been combined with physical data in order to evaluate social and environmental effects, e.g. employment, air pollution, and total energy requirements (Bullard et al. 1978). In view of this operational advantage, environmental footprint accounts have

[5]http://www.ghgprotocol.org/.

[6]This concept may had been influenced by the earlier works of Borgstrom (1969) who talked about 'ghost acreages' to describe the additional agriculture and fishing grounds that some countries where using to support their populations, and Vitousek et al. (1986) who estimated the human appropriation of the planet's net primary production, among others.

increasingly been adopting input–output frameworks (see Wiedmann et al. 2006; among others). Social footprints, however, have not yet explored this fully.[7] The reason for such lagging development may have been data inaccessibility. However, the emergence of national and international statistical databases, compiling information on production-related social issues (examples provided in the Methodology) makes this advance now feasible. The combination of available big data sets with MRIOA settings, such as the WIOD, EXIOBASE and Eora, sets the path to a new era of social footprint indicators that can keep up with the economic complexities of today (Tukker and Dietzenbacher 2013).

Given its rising popularity among researchers, policy-makers and the public as a measure of ecological burden, other environmental indicators adopted the term footprint, with the water footprint and the carbon footprint among the most popular ones.[8] Application of footprinting methodologies using IOA and multi-regional input-output analysis (MRIOA) includes water footprinting, carbon footprinting, materials footprinting, biodiversity footprinting and employment footprints (McBain et al. 2014). There have also been MRIOA studies on developing a 'footprint family' for ecological, water and carbon footprinting (Galli et al. 2012, 2013). Although the examples outlined above use input-output analysis, there exists numerous methodologies for conducting footprint calculations, and the lack of a single comparable methodology weakens the value of the tool. Academic debate regarding what constitutes a footprint and the validity of the numbers used continues. In particular, the key concerns about the footprinting methodology include translating environmental impacts into a single unit (such as global hectares) to enable comparison, the aggregation of different impacts and the hypothetical nature of using more than one planet (Blomqvist et al. 2013; van den Bergh and Grazi 2014). To allay these concerns, the methodology used in Chap. 2 does not seek to create one footprint number using a single unit or aggregation of impacts, but instead presents an array of indicators in their basic units of measurement as has been demonstrated with carbon and water footprints.

As demonstrated above, the term 'footprint' is applied to a number of different applications, without necessarily having consistency in methodology. In a recent speech, comparison was drawn between the evolution of the environmental movement and our environmental understanding with our knowledge of the finance and inequality:

> We can draw some parallels here with our expanding environmental consciousness… By comparison, the equivalent kind of awareness in the financial sector—the idea that private misbehavior can have a broader social cost—is only in its early stages. It is akin to the

---

[7]There has been, though, suggestions to combine Social Life Cycle Assessments with IOA, which would be akin to estimating commodity-level footprints of the given social issues evaluated.

[8]There are several definitions of 'carbon footprint', which differ significantly from the original 'carbon footprint' component of the 'ecological footprint' (GFN 2014).

initial period of environmental consciousness, which focused on the banning of lead from petroleum products. Just as we have a long way to go to reduce our carbon footprint, we have an even longer way to go to reduce our "financial footprint".[9]

The methodology behind the use of existing 'social footprints' and 'economic footprints' remains loosely defined and is rarely used (see Čuček et al. 2012). As a result of this heterogeneous collection of 'footprint indicators', up to now "there is no standard and clear definition of [what is] a 'footprint'" (Ibid, p. 10). Nevertheless, we may argue that a main aim of footprints is the accounting of *total* (direct and indirect) burdens linked to a given economic activity.[10] Such burdens are usually calculated for different economic commodities, and said to be embodied or embedded in them.[11] These embedded burdens are then aggregated so as to describe the impact of businesses or economic sectors, and following the aim of the original footprint concept, they are reflected upon the consumers of such commodities (e.g. consumer countries' footprints). Several applications of different methodologies have been made, including looking at the employment footprint of nations (Alsamawi et al. 2014); calculating the human footprint on biodiversity (Burton et al. 2014); using a human footprint map for species conservation (Di Marco et al. 2013), the footprint of culture and identity (Minnaert 2014) and using a social footprint to manage corporate sustainability (McElroy and van Engelen 2012). Likewise, Gómez-Paredes et al. (2014) have explored the use of detail labour data and IOA for the calculation of Indian commodities' labour footprints, as means to assess embodied violations of international labour rights and principles, along entire production chains.

If we examine the concept of a social footprint further, what is the maximum assimilative capacity of humans? Is it the limit of our population size (which is, in fact, renewable and changeable)? Is it the number of hours a day that a person can work (maximum of 24 or the ideal of eight) or lifespan (which again varies from country to country, and even between genders)? Or is it related to how a person can live life in a happy and just way? Although not a social footprint, the GFN suggests that we need to consider the ecological footprint of an individual (estimated to be 1.8 global hectares per person) combined with a Human Development Index (referred to above) of above 0.8 to have sustainable human development where people can live fulfilling lives without environmental degradation.[12] This contrasts with Bravo's review of the Human Sustainable Development Index which combines the three socio-economic aspects of the Human Development Index (life expectancy at

---

[9]*Economic Inclusion and Financial Integrity—an Address to the Conference on Inclusive Capitalism* by Christine Lagarde, Managing Director, International Monetary Fund, London, May 27, 2014. https://www.imf.org/external/np/speeches/2014/052714.htm.

[10]It must be noted that not all the so-called 'footprint indicators' fit this logic, e.g. the 'corruption footprint' or the 'job footprint' (for a brief description of these and other see Čuček et al. 2012).

[11]'Commodity footprints' are often the product of Life Cycle Assessments.

[12]http://www.footprintnetwork.org/en/index.php/GFN/page/fighting_poverty_our_human_development_initiative/. Accessed 3/7/14.

birth, education and income) with an environmental dimension of per capita CO_2 emissions (Bravo 2014).

The use of satellite accounts for special purposes in input-output analysis, such as measuring health or environmental impacts, is outlined in the System of National Accounts (United Nations Statistics Division 2008, Chap. 29). Different approaches to assessing social impacts using LCA-MRIO analysis have been used, such as by Benoit-Norris et al. (2011) and Rugani et al. (2012). Studies linking MRIOA and social impacts are emerging particularly in relation to inequality between global trading partners (Alsamawi et al. 2014; Prell et al. 2014) and assessing product supply chains using the GTAP MRIO database and associated Social Hotspot Database (Benoit-Norris et al. 2012). There are many benefits of using MRIOA for supply chain analysis. The ability to trace millions of supply chains throughout the world has come about through the painstaking creation of detailed MRIO databases based on national economic accounts. Accounts holding detailed social information as they relate to national economic accounts (social satellite accounts) are being developed through other data sources such as the World Bank, World Health Organisation, International Labour Organisation. To trace these supply chains through economic data without the use of MRIOA would be almost impossible, and certainly very time consuming. The scope provided by MRIOA is very large. Using economic data can provide rigor to the process of understanding the social impacts of global trade.

References

Alsamawi, A., Murray, J., & Lenzen, M. (2014). The employment footprints of nations uncovering master-servant relationships. *Journal of Industrial Ecology, 18*(1), 59–70.

Benoit-Norris, C., Cavan, D. A., & Norris, G. (2012). Identifying social impacts in product supply chains: Overview and application of the social hotspot database. *Sustainability, 4,* 1946–1965.

Benoit-Norris, C., Vickery-Niederman, G., Valdivia, S., Franze, J., Traverso, M., Ciroth, A., et al. (2011). Introducing the UNEP/SETAC methodological sheets for subcategories of social LCA. *International Journal of Life Cycle Assessment, 16*(7), 682–690.

Blomqvist, L., Brook, B. W., Ellis, E. C., Kareiva, P. M., Nordhaus, T., & Shellenberger, M. (2013). Does the shoe fit? Real versus imagined ecological footprints. *PLoS Biology, 11*(11), e1001700.

Bravo, G. (2014). The human sustainable development index: New calculations and a first critical analysis. *Ecological Indicators, 37,* 145–150.

Bullard, C. W., Penner, P. S., & Pilati, D. A. (1978). Net energy analysis—Handbook for combining process and input-output analysis. *Resources and Energy*, 1, 267–313.

Burton, A. C., Huggard, D., Bayne, E., Schieck, J., Solymos, P., Muhly, T., et al. (2014). A framework for adaptive monitoring of the cumulative effects of human footprint on biodiversity. *Environmental Monitoring and Assessment, 186*(6), 3605–3617.

Costanza, R., Kubiszewski, I., Giovannini, E., Lovins, H., McGlade, J., Pickett, K. E., et al. (2014). Time to leave GDP behind. *Nature* 505:283–285.

Čuček, L., Klemeš, J. J., & Kravanja, Z. (2012). A review of footprint analysis tools for monitoring impacts on sustainability. *Journal of Cleaner Production, 34,* 9–20.

Di Marco, M., Rondinini, C., Boitani, L., & Murray, K. A. (2013). Comparing multiple species distribution proxies and different quantifications of the human footprint map, implications for conservation. *Biological Conservation, 165,* 203–211.

Galli, A., Weinzettel, J., Cranston, G., & Ercin, E. (2013). A footprint family extended MRIO model to support Europe's transition to a one planet economy. *Science of the Total Environment, 461–462,* 813–818.

Galli, A., Wiedmann, T., Ercin, E., Knoblauch, D., Ewing, B., & Giljum, S. (2012). Integrating ecological, carbon and water footprint into a "footprint family" of indicators: Definition and role in tracking human pressure on the planet. *Ecological Indicators, 16,* 100–112.

GFN. (2014). *Footprint basics—Overview.* http://www.footprintnetwork.org/en/index.php/GFN/page/footprint_basics_overview/. Accessed 6 August 2014.

Gómez-Paredes, J., Yamasue, E., Okumura, H., & Ishihara, K. N. (2014). The labour footprint: An input–output based framework to assess labour dimensions of economic activity (submitted).

Kubiszewski, I., Costanza, R., Franco, C., Lawn, P., Talberth, J., Jackson, T., et al. (2013). Beyond GDP: Measuring and achieving global genuine progress. *Ecological Economics, 93,* 57–68.

Leontief, W. (1936). Quantitative input and output relations in the economic system of the United States. *The Review of Economic and Statistics, 18*, 105–125.

McBain, D., Wiedmann, T., & Lenzen, M. (2014). The contribution of input-output analysis to understanding impacts of consumption—A review update. *Journal of Industrial Ecology* (submitted).

McElroy, M. W., & van Engelen, J. M. L. (2012). *Corporate sustainability management: The art and science of managing non-financial performance.* Hoboken: Taylor and Francis.

Minnaert, T. (2014). Footprint or fingerprint: International cultural policy as identity policy. *International Journal of Cultural Policy, 20*(1), 99–113.

Pelletier, N., Ustaoglu, E., Benoit, C., & Norris, G. (2013). Sustainability in trade and development policy. A life cycle approach to understanding and managing social risk attributable to production and consumption in the EU-27. In J. R. C. I. f. E. (Ed.), *Sustainability.* Luxembourg: Publications Office of the European Union European Commission.

Prell, C., Feng, K., Sun, L., Geores, M., & Hubacek, K. (2014). The economic gains and environmental losses of US consumption: A world-systems and input-output approach. *Social Forces.*

Rees, W. E. (1992). Ecological footprints and appropriated carrying capacity: What urban economics leaves out. *Environment and Urbanization, 4*(2), 121–130.

Rees, W., & Wackernagel, M. (1996). Urban ecological footprints: Why cities cannot be sustainable—And why they are a key to sustainability. *Environmental Impact Assessment Review, 16*(4), 223–248.

Rugani, B., Panasiuk, D., & Benetto, E. (2012). An input–output based framework to evaluate human labour in life cycle assessment. *The International Journal of Life Cycle Assessment, 17*(6), 795–812.

Tukker, A., & Dietzenbacher, E. (2013). Global multiregional input-output frameworks: An introduction and outlook. *Economic Systems Research, 25*(1), 1–19.

United Nations Environment Program. (2009). *Guidelines for social life cycle assessment of products.* Milan: United Nations.

United Nations Statistics Division. (2008). *The system of national accounts (SNA).* New York: United Nations.

Ura, K., Alkire, S., Zangmo, T., & Wangdi, K. (2012). *A short guide to gross national happiness index.* Bhutan: The Centre for Bhutan Studies.

van den Bergh, J. C. J. M., & Grazi, F. (2014). Ecological footprint policy? Land use as an environmental indicator. *Journal of Industrial Ecology, 18*(1), 10–19.

Wackernagel, M., & Rees, W. (1996). *Our ecological footprint: Reducing human impact on the earth.* Gabriola Island, British Columbia, Canada: New Society Publishers.

Wiedmann, T., et al. (2006). Allocating ecological footprints to final consumption categories with input–output analysis. *Ecological Economics*, *56*(1), pp. 28–48. Available at: http://linkinghub.elsevier.com/retrieve/pii/S0921800905002600. Accessed October 27, 2012.

Chapter 6
A Social Footprint of Nations: A Comparative Study of the Social Impact of Work

6.1 Introduction

Writing in 1910, British journalist and author Arnold Bennett spoke about the newly emerged white collar workforce who worked eight hours a day, five days a week and generally found their work to be something to be endured rather than enjoyed. Taking the case of a Londoner working in an office by way of example, Bennett (1910) observes of the work day:

> He persists in looking upon those hours from 10 to 6 as 'the day', to which the 10 hours preceding them and the six hours following them are nothing but a prologue and epilogue.... If a man makes two-thirds of his existence subservient to one-third, for which admittedly he has no absolutely feverish zest, how can he hope to live fully and completely? He cannot.[1]

Work is essential for most people to live a full and complete life. It can be alternatively a source of income, education, friendship, productivity, purpose, fulfilment and even joy. Whilst some might have 'no absolutely feverish zest' for work, studies have shown that full time work results in better health than part time or no work, for example (Frech and Damaske 2012; Zheng and Land 2012). National governments such as that of the UK have even based public health strategies on the link between work, health and wellbeing (Black 2012). The 2014 International Labour Organisation (ILO) *World at Work* report found that countries that made the greatest investment in quality jobs found the greatest improvement in living standards as compared to developing countries that did not focus attention to the quality of jobs despite what we know about the benefits of quality employment. However, far from enjoying work, many people find work places them in vulnerable and even life threatening positions. More than half of the developing world's workers (approximately 1.5 billion people) are classified as being in vulnerable employment, trapped

---

[1]Arnold Bennet, *How to Live on 24 Hours a Day*, Chapter IV. http://www.gutenberg.org/ebooks/2274 Accessed 5/6/14.

A. Alsamawi et al., *The Social Footprints of Global Trade*,
Environmental Footprints and Eco-design of Products and Processes,
DOI 10.1007/978-981-10-4137-2_6

in a cycle where low incomes limit the ability to invest in family and future generations' health and education. Of these, 839 million people in developing countries are classified as living in working poverty, where they earn less than US$2 per day (International Labour Organisation 2014b). There is clearly an argument for work, and quality work in particular, benefitting people. However increasingly we are hearing more about the dark side of work—excessive work hours, unsafe working conditions, bonded labour, child labour, human trafficking and gross inequality. In fact, some of these issues have become so prevalent in working conditions as to have been named modern slavery. The ILO estimates that up to 21 million people worldwide are victims of forced labour, which generates US$150 billion profits in the private economy each year (International Labour Organisation 2014a). Whilst slavery was outlawed by different nations at different times and at different levels, often the introduction of Slave Trade Acts (in 1788 in the UK; and in 1794 in the United States) is considered as the starting point for the abolition of slavery. However, the act of working for the benefit purely of others and not the individual has persisted. Alsamawi et al. (2014) created employment 'footprints', showing that the citizens of some nations work primarily to support the lifestyle of those living in 'Master' nations, creating a master-servant relationship between nations on the basis of international trade. Whilst being recognised as a significant issue, governments have been slow to respond. In 2014 a bill was put before the UK parliament to eradicate modern slavery in UK operations, although at the time of writing the changes to the Companies Act would not include supply chains extending beyond the UK. In 2016 the European Union is likely to enact new laws enforcing companies to report on human rights in their business relationships (Mason 2014).

Why should we care? Other than personal concern, the impact of business on people is also affecting inequality. The charity Oxfam released a report on global inequality in 2014, attributing the richest 85 people in the world with holding as much wealth as the bottom half of the world's population (Fuentes-Nieva and Galasso 2014). When speaking about the concept of inclusive capitalism, Christine LaGarde, the Head of the International Monetary Fund, said:

> A greater concentration of wealth could—if unchecked—even undermine the principles of meritocracy and democracy. It could undermine the principle of equal rights proclaimed in the 1948 Universal Declaration of Human Rights.[2]

French economist Piketty (2014) argues in his book, *Capital in the 21st Century* that there is rising inequality in the global distribution of wealth, based on his study of wealth and income data from the USA, Britain and France. The book focuses on the role of work and income to generate wealth versus the rate of return on capital. Piketty concludes that the trend towards inequality is unlikely to reverse based on current economic policies adopted across the globe. Published in French in 2013 and English in 2014, the book shot to the bestsellers list in many countries.

---

[2]Economic Inclusion and Financial Integrity—an Address to the Conference on Inclusive Capitalism by Christine Lagarde, Managing Director, International Monetary Fund, London, May 27, 2014. https://www.imf.org/external/np/speeches/2014/052714.htm.

Work, inequality and social impact are becoming key interests for social researchers, economists, politicians, policy makers and even the general public.

The social contract[3] is being challenged. This is not surprising given that it relies on a seventeenth century understanding that the self-interest of rational man ensures we willingly exchange some of our freedoms in order to enjoy the benefits of whatever political and economic system we, or rather rational man, has tacitly agreed to (Hobbes, in Macpherson 1985). Rational man is coming under scrutiny, not only on gender equity grounds but also on racial grounds (Friend 2004). The premise of this tacit agreement between Western governments and their elite, mainly male power-brokers is being questioned. The pact wasn't designed to include women, the poor and the disenfranchised, all of whom were not considered capable of higher-order (rational) thinking, no matter where in the world they lived. This exclusion from higher-order thinking also had repercussions for what could be considered an individual's needs. Maslow's hierarchy of needs implied that those suffering such inequality that their basic needs were barely satisfied were incapable of higher order thinking thus legitimising the distribution of power in favour of the already powerful (Jackson et al. 2004). Echoes of this still linger. Now it seems the needs of the developed world are consuming the outputs of the developing world. Globalisation has proved unable to tackle inequality and in some cases has exacerbated it (International Monetary Fund 2007; Dreher and Gaston 2008). The needs of vulnerable workers in the developing world seem to be stuck at the bottom of Maslow's hierarchy unseen and unheard by some of the world's largest corporations who knowingly or unknowingly exploit them (Bangladesh All Party Parliamentary Group 2013). Now, thanks in some part to the power of the media their voices are beginning to be heard. Our research supports those who work to amplify the sound and provoke action on behalf of workers everywhere who suffer gross inequality.

In this chapter we look at the social footprint of labour for 20 countries. We present an approach to compiling a suite of indictors as a social footprint, measuring eight dimensions of social impact for the average worker, and comparing this across the globe. We use the MRIOA approach to footprinting, and further address questions around this use of footprinting methodology such as additively and proportionality in the methodology section.

6.2 Background

All goods and services consumed have a social input. Just as Leontief found that producing a car had an input of pollution, producing a car also has an input of labour. Sometimes that input is local or domestic (the person who made your cup of

[3]Social contract theory was developed by Hobbes in the first half of the seventeenth century. It is regarded by some as the basis of Corporate Social Responsibility (UNEP 2009).

coffee this morning, for example, who is living and working in the same country as you) and sometimes the labour is imported or embedded within a product (the person who grew the coffee in another country). An ILO report into the World of Work in 2014 found that, "*… economic growth is not sustainable when it is based on poor and unsafe working conditions, suppressed wages and rising working poverty and inequalities.*" (International Labour Organisation 2014b). This chapter presents a social footprint of consumption for 10 developed and 10 developing nations. The consumption consists of both the labour inputs and impacts of people working for domestic consumption, and the imported labour inputs and impacts through the consumption of goods produced internationally. This is similar to how a carbon footprint works, where a carbon footprint looks at domestic production of carbon dioxide emissions, plus those imported to a country embodied in goods and services for local consumption.

6.3 Methodology

In this chapter, the global social footprints of labour are calculated by using MRIO tables. Basic input output theory and extended input-output analysis, as applied in this case study, are described in the paper by Alsamawi et al. (2014) and will not be repeated here for want of space. This study extends previous work on calculating some of the indirect social impacts of nations (Alsamawi et al. 2014; McBain and Alsamawi 2014) by combining multiple satellite accounts to develop a more complex and complete social footprint of work. We created nine satellite accounts, $\mathbf{Q}_{\mathbf{x}}$, which represent the economy wide social accounting identity. The analysis was run using 187 countries and over 15,000 sectors from the EORA MRIO database (Lenzen et al. 2013) using the Leontief demand-pull model, which interprets $\mathbf{Q}$ as the total social inputs needed to fulfil the final demand, or the social footprint. The satellite accounts are employment ($\mathbf{Q}_{\mathbf{emp}}$) and income ($\mathbf{Q}_{\mathbf{inc}}$) as calculated in the footprints in (Alsamawi et al. 2014), fatal accidents ($\mathbf{Q}_{\mathbf{fa}}$), non-fatal accidents ($\mathbf{Q}_{\mathbf{nfa}}$), days lost ($\mathbf{Q}_{\mathbf{dl}}$) and wages lost ($\mathbf{Q}_{\mathbf{w}}$), for economic sectors as calculated in health and safety footprints in, inequality ($\mathbf{Q}_{\mathbf{ineq}}$) as calculated in the inequality footprint in (see Chap. 8). In addition, we used the same method as that used in (Alsamawi et al. 2014) to calculate for the first time the unemployment ($\mathbf{Q}_{\mathbf{unemp}}$) and strikes and lockouts ($\mathbf{Q}_{\mathbf{str}}$) satellite accounts.

Various databases from different resources were incorporated into this work making use of MRIOA's ability to work with big data. Our data were based on the International Labour Organisation's LABORSTA database (ILO 2012); the United Nations System of National Account UNSNA-Official Country database (UNSD 2011) for employment data; the Standardized World Income Inequality Database (SWIID) version 3.1 (Solt 2009); the OCED database (http://stats.oecd.org/) and the Gini index database from the World Bank (http://data.worldbank.org/) for before and after tax income and inequality data; the Asean Occupational Safety and Health Network (http://www.aseanoshnet.org/) for accidents, strikes and lockouts; and the Eora Multi-Regional Input-Output (MRIO) database (Eora 2012; Lenzen et al. 2012a, b) for global inter-industry transactions data.

6.4 Limitations of the MRIO Methodology for Footprinting

Using static input-output techniques for footprinting exercises requires making a number of assumptions. Firstly, the constancy of the coefficients in the direct requirements and final demand matrices implies a fixed production and consumption recipe; prices have no influence on producers and consumers alike in choosing their purchases. Secondly, and this is a consequence of the fixed production recipe, Leontief's traditional demand-pull model of the economy assumes proportionality: a doubling of consumption will require a doubling of production, there are no economies of scale, there is no slack in production factors. For more details, see (Miller and Blair 2010). Ultimately, this means that strictly speaking, static IO multipliers and footprints derived thereof have to be interpreted in an ex-post perspective, i.e. as relationships that have played out in the past, and that do not necessarily imply the same cause-and-effect chains in the future. In order to enable future projections, models have traditionally gone beyond the simple static IO mechanism, for example in choice-of-technology models (Duchin and Levine 2011), or the widespread computable general equilibrium (CGE) models (Rose 1995). LCA practitioners are also moving towards incorporating more dynamic effects, for example in consequential LCA (Earles and Halog 2011; Finnveden et al. 2009; Weidema 1993).

Another methodological issue in social footprinting is the selection of indicators. Social indicator databases include estimates on gender participation, literacy and health, that are not readily fit for integration into an (MR)IO framework. This is because first, some of those indicators do not allow allocation of values across impact-causing industries and second, respective quantity is not additive. The latter point means that it does not make sense to add percentages of literacy to percentages of child deaths, one can only add absolute numbers. This requirement is important in IO analysis, because the Leontief inversion essentially adds up contributions from numerous supply chains. Therefore, only additive quantities can be integrated.

There are limitations to the use of MRIOA for footprinting or global supply chain analysis, which should be understood for the interpretation of results. Limitations of input-output analysis are well understood. For example, if a business understands its size in proportion to the industry sector, it can estimate responsibility proportionally for the chosen indicator. However, if a business does not behave in an average manner (e.g. it uses only recycled material inputs to production) the results will need to be adjusted accordingly. IOA is based on macroeconomic data, collected in accordance with the SNA (United Nations Statistics Division 2009). A nation's accounts depend on that nation being able to accurately collate data relating to labour, employment, income, taxes, trade and even population. However, where a country has a significant underground economy or a population employed in non-paid work (e.g. domestic labour, subsistence farming), the SNA is unable to account for that economic activity (Waring 1988). Further discussions on IO analysis limitations and strengths in general such as sector averages, uncertainties of data, scope and double counting can be found in texts such as by Murray and Wood (2010). Further details regarding the data quality of the Eora MRIO tables, including data optimisation, standard deviation settings and the confidence of UN Main Aggregates and Official Country Data and UN Comtrade data can be found in Lenzen et al. (2012b).

6.5 Results

The study analysed data from 187 countries, across 9 indicators—Employment (people × millions), Income (US$ × billions), Fatal workplace accidents (people × thousands), Non-fatal workplace accidents (people × millions), Days Lost due to workplace injury (days × millions), Wages Lost due to workplace injury (US$ × millions), Unemployment (people × millions) and Strikes and Lockouts (number of events). Figures 6.1 and 6.2a, b illustrate the footprint results.

The results show that there is a similarity in the social footprint of work in developed nations, but there is a great variability in the footprint for developing nations. The countries shown in Fig. 6.1 were chosen out of the 187 countries analysed as representatives for developed and developing nations. On the right hand side of Fig. 6.1 we display the countries with developed economies. In general, these countries import or consume as much labour as they use domestically, which is why the two lines (red and blue) are very close in size and shape. This graph highlights that the social impacts of consumption in these countries are shared almost equally between the residents of the developed country and by producers in non-resident countries (often developing countries). By contrast, the developing countries shown on the left hand side have larger domestic footprints than import footprints because they represent the world's producers. This relationship was explained in Alsamawi et al.'s paper on trade Master and Servant relationships (Alsamawi et al. 2014). The countries on the left hand side of Fig. 6.1 also represent the rapidly developing BRIC countries (Brazil, Russia, India and China). Madagascar was included as a developing country because 70% of the population works for the export market (Alsamawi et al. 2014) and hence it has a small domestic footprint but contributes significantly to the global imports footprints, particularly for countries like France.

Figure 6.2a, b represent the total social footprint of work for developing (a) and developed (b) nations. These plots shows a multiplier for a given country (m_i) that has been normalised to its final demand

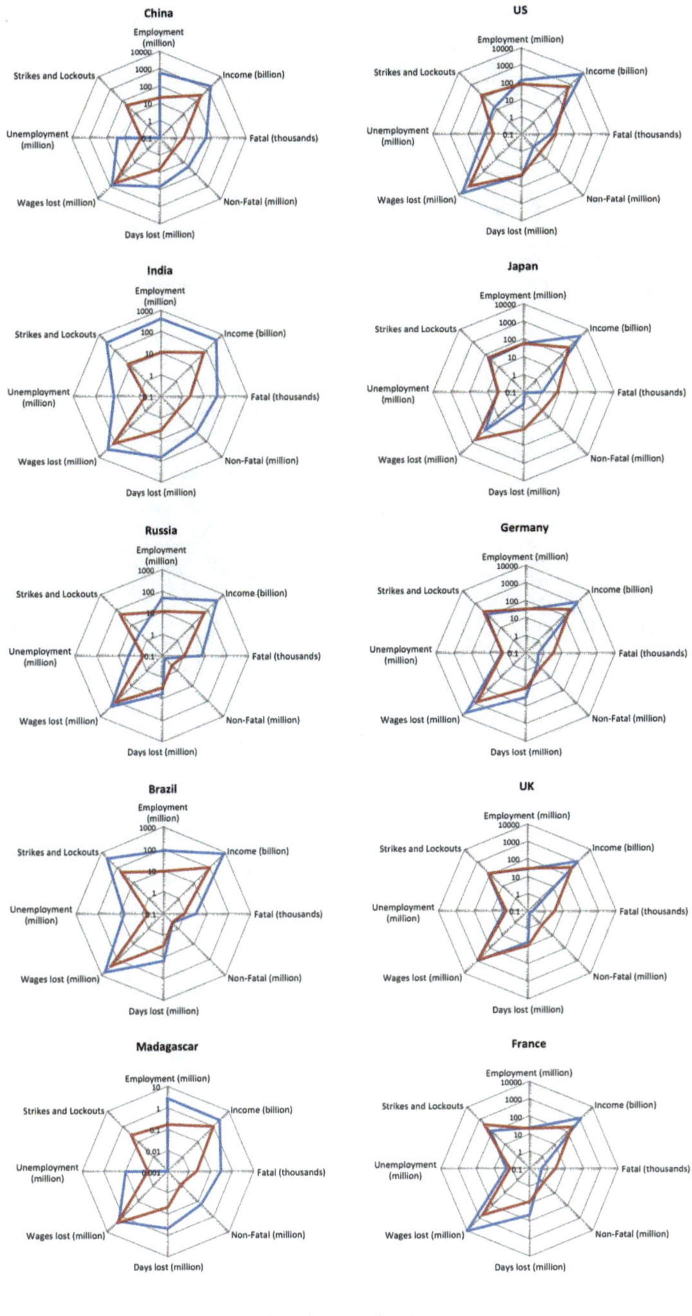

Fig. 6.1 The social footprint of work: domestic and imported social footprint

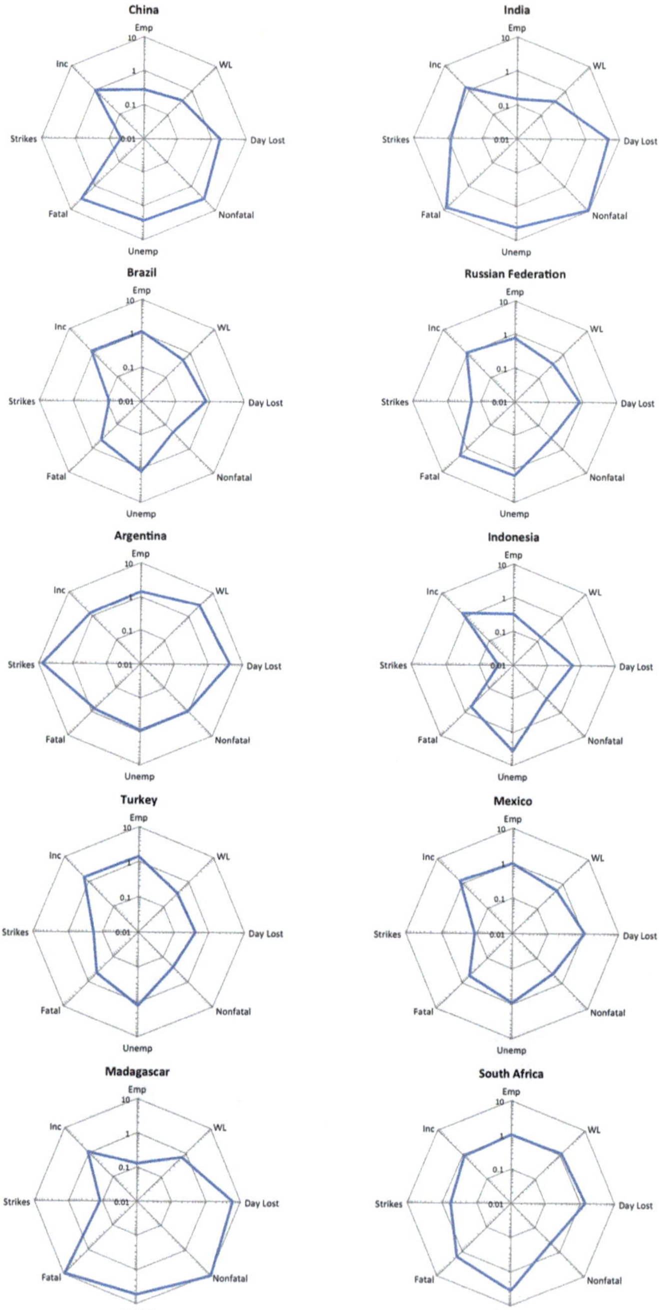

Fig. 6.2 **a** Social impact of work: total social footprint of work for developing nations. **b** Social impact of work: total social footprint of work for developed nations

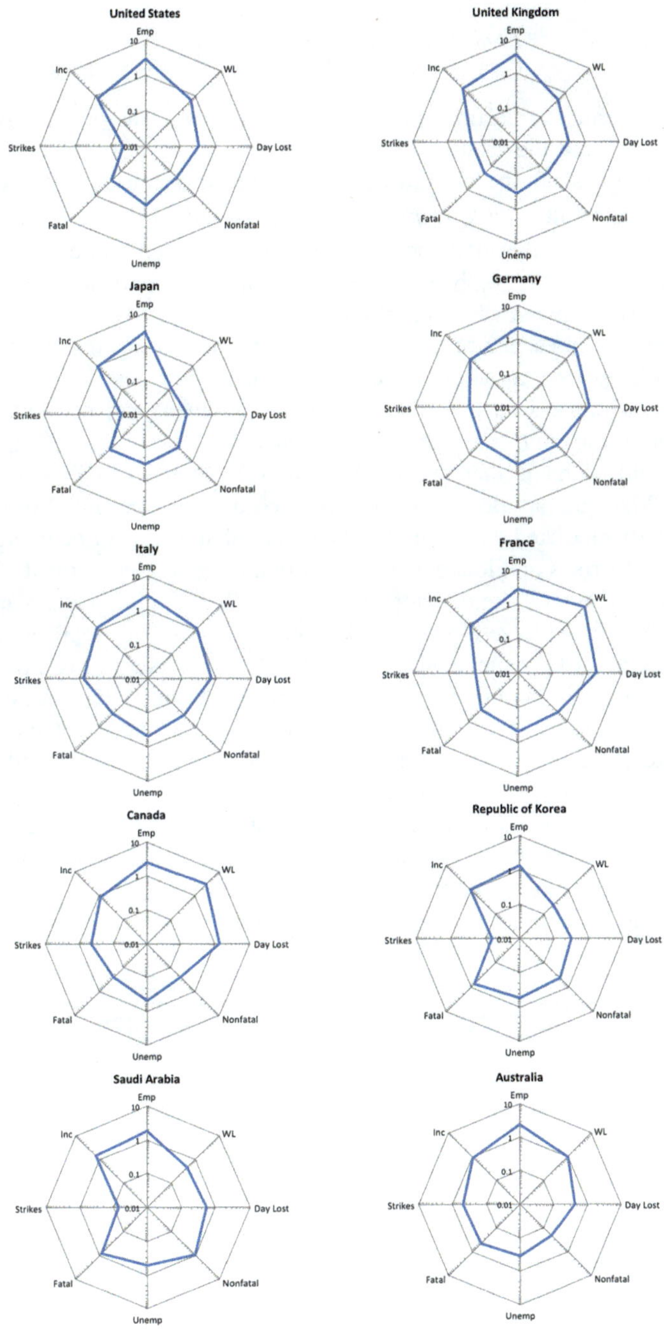

Fig. 6.2 (continued)

$$(FD_i), \quad m_i = \left(\frac{FPi}{FDi}\right) \quad \text{where } FPi$$

is the total footprint (domestic + import) of the country *i.* The polygon represented in Fig. 6.2a, b show the relative total social footprint of work, where a footprint equal to one represents the worldwide average of all countries. Otherwise stated, the ratio (or the results) $\left({}^{m_i}/_{\bar{m}}\right)$, $\bar{m}$ is the world average. In these figures, both the size and the shape of the total footprints can be compared and contrasted. In the developed countries (Fig. 6.3b) we see a broad similarity of shape, with a positive social impact skew towards high employment, income and wages lost (where wages lost represents the presence of compensation for sickness or injury). In the developing countries we see a variety of footprint shapes, with a general skew towards negative social impacts such as fatalities, unemployment and non-fatal accidents (for example, see China, India and Madagascar).

Note should be made that two of the countries depicted in Figs. 6.1 and 6.2, China and Madagascar, do not have any recorded Strikes and Lockouts data. Although an extrapolated data point has been included, it cannot be interpreted as a low number of strikes and lockouts (whether due to good employment laws or bad recording of figures). Of the developed countries, Republic of Korea, Japan, Saudi Arabia and the US all have dips for strikes and lockouts. This represents a lack of worker rights to strike rather than a low rate of strikes due to high worker satisfaction. An interesting contrast exists with Argentina and India, where there is a relatively high number of Strikes and Lockouts, higher than all of the developed countries excepting Italy. This represents the worker ability to strike and the ability to lobby for better pay or conditions.

The industry sectors contributing to the footprints vary significantly in their impact shapes. For example, agriculture is a significant component of the export market from the developing countries and contributes to the domestic footprint, but contributes more significantly to the imports footprint of developed countries.

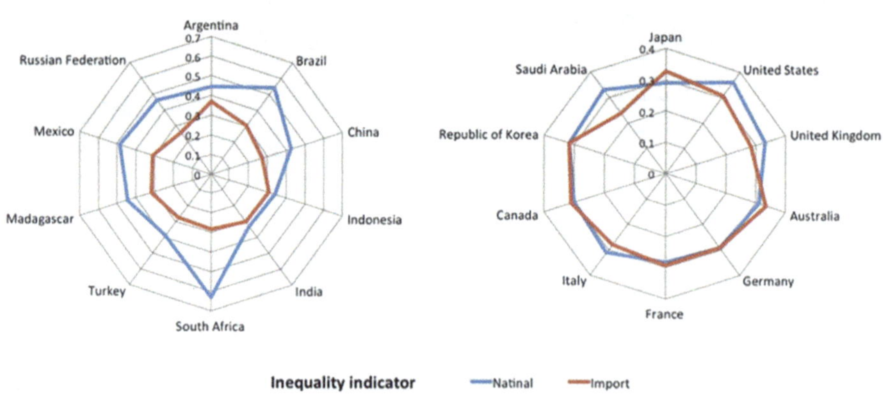

Fig. 6.3 The inequality of nations footprints for developing (a) and developed (b) countries

This was demonstrated using structural path analysis in (Alsamawi et al. 2014, see Table 1) for trade paths from developed to developing countries. In the developing countries agriculture is generally high in employment, fatalities, non-fatal injuries and unemployment but low in strikes and lockouts and low in income. This is representative of the high amount of non-unionised labour used in the agriculture field in developing countries and the number of small operations. By contrast in developed countries agriculture is often operated at an industrial scale. Mining, in developing countries, can be similarly represented by high employment, fatalities and injuries however in developed countries mining is represented by high strikes and lockouts, high injuries and fatalities, high wages but a low percentage of employment. From a social-LCA perspective, the social impact of the agriculture sector is demonstrated in Chap. 5, Fig. 5.1 (Social Hotspots for Developing Countries in Agricultural Sectors). Although the category of Labour and Decent work was chosen to complement the social footprint developed in this chapter, many more details derived through social LCA could be analysed and used to interpret the MRIOA social footprint. By combining the detailed analysis presented by social LCA in Chap. 5, Fig. 6.1 with the domestic and imported impact analysis developed as a footprint here, a more thorough picture of social impacts on the ground and due to international trade can be developed.

Legislation also plays a large part in the shape of the domestic footprint and the subsequent total footprint (Figs. 6.1, 6.2 and 6.3). European countries such as France, for example, have labour laws which support safe working conditions, decent remuneration and work life balance to be achieved through a 35 h working week.[4] They also have taxes and transfer payments which promote equality within society (see Fig. 6.2b). The European countries represented in Fig. 6.2b (France, Italy, Germany) all have a similar shaped footprint, as would be expected from operating under similar legislation with respect to working rights (note that the UK has opt out provisions for the Working Time Directive, which may be one reason why it has a slightly dissimilar footprint to other European countries, as well as other differences in legislation and the economy). The social footprints of work for these countries are framed almost around a triangle of high income, high numbers of days lost (representing high standards for sick leave provisions) and high strikes and lockouts (representing a workforce with the right to strike). These high social standards are to an extent passed on through procurement and purchasing conditions. Japan and the US have similarly shaped footprints (see Fig. 6.2b), with more fatalities and unemployment that the European countries. We can see from Fig. 6.2 that the inequality footprint (developed from GINI coefficients, where 0 represents perfect equality and 1 represents perfect inequality) of imports is proportionally larger in developed countries (b) than developing countries (a). The embedded social impacts are roughly divided equally between imports and exports for

[4]For further information see the European Working Time Directive 2003/88/EC of the European Parliament and of the Council of 4 November 2003 concerning certain aspects of the organisation of working time http://eur-lex.europa.eu/LexUriServ/LexUriServ.do?uri=CELEX:32003L0088:EN:HTML, http://www.ilo.org/global/statistics-and-databases/lang--en/index.htm.

developed countries (see Fig. 6.2b). However, for developing countries, who consume much less, the social footprint of imports is much smaller than the domestic social footprint. In developing countries there is more inequality domestically than what is embodied in imports. The inequality footprints based on the Gini Coefficients correlate closely with the domestic and imported social footprints calculated for Fig. 6.1.

6.6 Discussion

6.6.1 Role of Governments

The inequality inherent within the social footprints of work is integral to economic growth and prosperity and yet can hinder development. Weak labour laws and wage setting mechanisms, controlled by governments through legislation, have led to social inequalities in trade. Through organisations such as the IMF and the ILO there is increasing recognition of and pressure on governments to effectively govern labour regulations and markets to reduce inequality. This is particularly relevant when weak employment regulations lead to low participation in formal employment. When considering a footprint of labour based on trade, we must be mindful of the impact of the black market. The informal or black market economy is not, by its very nature, captured by the System of National Accounts. In some labour markets, over 50% of people work for the informal economy. For example, it is estimated that 68% of Indonesia's workforce work for the informal market.[5] As stated previously, there is an estimated 21 million people worldwide working as slave labour and this labour generates approximately US$150 billion profits in the private economy each year (International Labour Organisation 2014a). This is not captured within our social footprint of work. Some value will be captured at the point in the supply chain where the good or service is sold or purchased in the formal economy, but the labour will not be reflected. Using other methods of analysis within the MRIOA sphere, such as structural path analysis coupled with sector disaggregation using LCA data, can bring some clarity to black market trade through the use of more detailed sector data where informal or black market trade is known to occur (see for example Moran et al. 2014). This method, often referred to as LCA-MRIO analysis, can bring definition to our understanding of social impacts contained within global supply chains and what is happening on-the ground. In action, examples such as work in Argentina show that government regulation can bring about higher participation in the formal labour

[5]"According to the Central Bureau of Statistics, 68% of Indonesians was employed in informal economy in 2009, often with low pay, hazardous working conditions and no social security". Source: http://www.ilo.org/jakarta/areasofwork/informal-economy/lang–en/index.htm. Accessed 19/08/14.

market, improved social protection and reduce inequality whilst improving business competitiveness (International Labour Organisation 2014b).

The quality of data is an important part of analysis of the informal labour market regardless of the method of analysis used, and developing more robust data on trade and informal markets is an essential part of the process to recognise the scale of labour issues relating to work. Calls for the development of a system of socio-economic accounts (McBain and Alsamawi 2014) go some way towards this, however the collection and collation of accurate data by governments is also essential for addressing inequality, the social impact of work, and the structure of national economies. The social footprint of work, as demonstrated here, will provide an effective tool for comparing progress both temporally within countries and across countries. Social footprint results can be used for communicating within government and to other interested stakeholders of government such as international agencies and NGOs, and businesses importing labour from particular countries.

6.6.2 Role of Business and Institutional Investors

Given the significance of the import component of the social footprint, the role of multinational enterprises in their sourcing policies and the role of institutional investors must also be considered. Legislation such as SEC1520, introduced under the US Dodd-Frank Act 2010 shows that governments do have the power to require business to understand the labour impacts associated with their materials sourcing, and that without tools such as demonstrated here, this information can be difficult to come by (King 2014). In the absence of strong local labour laws, purchasing codes and sourcing policies can give workers an opportunity for fair work conditions beyond national legislation. For example Fairphone, a social enterprise based in the Netherlands, has established secure supply chains to develop and deliver the first smartphones to the market that have high social standards embedded within the product. Working in the electronics industry, an industry sector associated with poor labour standards from metals mining through to manufacturing in developing countries, Fairphone aims to deliver safe conditions, fair wages, worker representation and employee wellbeing in the process of delivering their product (Fairphone 2014).

Although securing supply chains to high labour standards to this extent is unusual, many multinational enterprises have specific sourcing policies relating to human rights and working conditions. Unilever, ranked as the world's most sustainable company in a recent GlobeScan survey, has a long history of caring for its workers. This ranges from provision of housing to factory workers and schooling, healthcare and decent wages for plantation workers in the Congo over a century ago, through to current responsible sourcing policies. As part of their Sustainable Living Plan, Unilever aims to reduce their environmental footprint and increase their positive social impact. A social footprint could be a way to demonstrate their positive social impact (Economist 2014). In addition to purchasing decisions, institutional investment also plays a vital role in sending market signals. Fiona Reynolds, managing director at the UN-backed Principles for Responsible Investment, is quoted as saying: "*Labour standards in the supply chains of global food and beverage companies, including retailers, are an area of intense scrutiny for institutional investors. Our signatories understand the importance of this issue on the performance of their investments and the impact it has on the health and wellbeing of workers.*"[6]

6.6.3 Role of Consumers

When discussing social impacts and labour, the role of consumers must also be considered. Individual workers are not separated from the impacts these social footprints of work have shown—they are involved both as individuals that make up the sum and also as the driving forces for consumption. Consumers all, to differing extents, consume domestic and imported labour. The developed world, as shown in this chapter and in various journal articles (for example Alsamawi et al. 2014; Gómez-Paredes et al. 2014) consumes more than the developing world. Studies such as Moran et al. (2014) demonstrate that it is possible to allocate responsibility for social impacts along a supply chain, from production to final consumption. If, as a society, we are to move towards being more sustainable, understanding our impact on the world's society is surely as important as our impact on the world's environment. Just as environmental footprints have become a useful tool for communicating our use of planetary resources to consumers, business and governments, the social footprint has the potential to become a useful tool to communicate our unequal consumption of global labour. The social footprint can be used to highlight where labour is being sourced from, and what the main concerns for each country are in terms of social impacts as well as whether labour is for the domestic or international market.

6.7 Conclusion

It is said that "[t]he journey towards sustainability finds sustainable production and consumption at its very heart" (United Nations Environment Program 2009, p. 5). Additionally, it is widely accepted that sustainability encompasses social as well as environmental aspects (United Nations 2012). Hence, recognising environmental and social issues along supply chains, linking producers with final consumers, becomes crucial in any comprehensive sustainability assessment. This capacity, however, is challenged by the complexities of worldwide production and trade, which are

[6]http://www.theguardian.com/global-development/2014/jun/12/ethical-money-investmentfunds.

predicted to increase. For instance, after the 2008 financial crisis global trade volumes have grown steadily and with the "import demand of high-income economies doubling, [...it] is expected to accelerate" (World Bank 2014, p. 16). Furthermore, hosting the view that global trade is the key to economic growth and job creation, in the 2014 G20 summit "leaders delivered a strong statement of commitment to promote free trade" (B20 Coalition 2014, p. 2) resulting in the establishment of a 'Transatlantic Trade and Investment Partnership' (TTIP) and a 'Transpacific Partnership' (TPP), "two mega-PTAs [Preferential Trade Agreements]" (Ibid, p. 16).

As previously mentioned, MRIO models are suitable for the analysis of trade scenarios, and this chapter demonstrates that they are particularly useful for examining the social impacts of global trade. MRIOA may be employed in sustainability assessments of specific supply chains; or in a broader context, used to analyse the footprint of nations in a form of quantified sustainability assessment. Such analyses will require, among other considerations, the evaluation of environmental and social footprints (e.g. the CO_2 emissions and social impacts embodied in commodities traded). Then, embodiments in international trade may be traced in order to evaluate final consumers' footprints, as well as to assess global disparities in terms of ecological demands and social demands.

Research into combining several environmental footprint indicators using MRIOA is beginning to occur (see Galli et al. 2012). However, the social dimension of these composite indicator frameworks is still absent. It follows that if comprehensive multi-criteria sustainability assessments are going to be possible, IO-compatible social footprints are an essential and urgent extension. The array of social impacts that social footprints may measure, such as unemployment, occupational health and safety, and income inequality is demonstrated within this paper. This paper provides a methodology and example of a comprehensive social impact footprint for global trade. On its own, this is an advance in the field of social footprinting and contributes to our understanding of both the issues and communication of complex messages. It also highlights the need for further work in considering other major social impacts of our time embedded in global trade, such as gender inequality, discrimination, indigenous rights and living poverty. This research demonstrates the need for accurate country data, which is particularly essential when trying to capture the social impacts of the informal economy. This is particularly relevant for regions where economic activity is expected to increase. This not only includes current emerging economies (e.g. the BRIC nations), but those regions where greater production is projected. The use of combined LCA-MRIO analysis may provide some solutions.

The social footprint shows the social impacts of work for all countries, and in particular the differences in social impacts between developed and developing countries, and the labour embedded in imports and exports. If, as a society, we want to reduce inequality, we need to understand where it currently resides, what the drivers are, and what the ideal social footprint might look like for a country which promotes just working conditions for its own workers and those abroad working on its behalf. The social footprint of nations goes some way to demonstrating what the current social impact of work is, and how we can measure our progress as a society.

References

Alsamawi, A., Murray, J., & Lenzen, M. (2014). The employment footprints of nations uncovering master-servant relationships. *Journal of Industrial Ecology, 18*(1), 59–70.

Alsamawi, A., Murray, J., & Lenzen, M. (2016). The working conditions footprints of nations. *Journal of Cleaner Production* (submitted).

Bangladesh All Party Parliamentary Group. (2013). *After Rana Plaza: A report into the readymade garment industry in Bangladesh* (p. 60). Parliamentary, UK Parliament, London: Parlimentary Liaison Office.

Bennett, A. (1910). *How to live on twenty-four hours a day: Project Gutenburg*.

Benoit-Norris, C., Cavan, D. A., & Norris, G. (2012). Identifying social impacts in product supply chains: Overview and application of the social hotspot database. *Sustainability, 4,* 1946–1965.

Benoit-Norris, C., Vickery-Niederman, G., Valdivia, S., Franze, J., Traverso, M., Ciroth, A., et al. (2011). Introducing the UNEP/SETAC methodological sheets for subcategories of social LCA. *International Journal of Life Cycle Assessment, 16*(7), 682–690.

Black, D. C. (2012). Work, health and wellbeing. *Safety and Health at Work, 3*(4), 241–242.

Blomqvist, L., Brook, B. W., Ellis, E. C., Kareiva, P. M., Nordhaus, T., & Shellenberger, M. (2013). Does the shoe fit? Real versus imagined ecological footprints. *PLoS Biology, 11*(11), e1001700.

Bravo, G. (2014). The Human Sustainable Development Index: New calculations and a first critical analysis. *Ecological Indicators, 37,* 145–150.

Burton, A. C., Huggard, D., Bayne, E., Schieck, J., Solymos, P., Muhly, T., et al. (2014). A framework for adaptive monitoring of the cumulative effects of human footprint on biodiversity. *Environmental Monitoring and Assessment, 186*(6), 3605–3617.

Coalition, B. (2014). *Trade for growth a new trade regime for the 21st century*.

Costanza, R., Kubiszewski, I., Giovannini, E., Lovins, H., McGlade, J., Pickett, K. E., et al. (2014). Time to leave GDP behind. *Nature, 505*(16 January), 283–285.

Čuček, L., Klemeš, J. J., & Kravanja, Z. (2012). A review of footprint analysis tools for monitoring impacts on sustainability. *Journal of Cleaner Production, 34,* 9–20.

Di Marco, M., Rondinini, C., Boitani, L., & Murray, K. A. (2013). Comparing multiple species distribution proxies and different quantifications of the human footprint map, implications for conservation. *Biological Conservation, 165,* 203–211.

Dreher, A., & Gaston, N. (2008). Has globalisation increased inequality? *Review of International Economics, 16*(3), 516–536.

Duchin, F., & Levine, S. H. (2011). Sectors may use multiple technologies simultaneously: The rectangular choice-of-technology model with binding factor constraints. *Economic Systems Research, 23*(3), 281–302.

Earles, J. M., & Halog, A. (2011). Consequential life cycle assessment: A review. *International Journal of Life Cycle Assessment, 16,* 445–453.

Economist, T. (2014). *In search of the good business*. http://www.economist.com/news/business/21611103-second-time-its-120-year-history-unilever-trying-redefine-what-it-means-be. Accessed August 12, 2014.

Eora. (2012). *The Eora global multi-region input-output tables*. www.worldmrio.com. Sydney, Australia: ISA, The University of Sydney.

Fairphone. (2014). *Manufacturing—Putting employee wellbeing at front and centre*. https://www.fairphone.com/roadmap/manufacturing/. Accessed August 10, 2014.

Finnveden, G., Hauschild, M. Z., Ekvall, T., Guinée, J., Heijungs, R., Hellweg, S., et al. (2009). Recent developments in life cycle assessment. *Journal of Environmental Management, 91,* 1–21.

Frech, A., & Damaske, S. (2012). The Relationships between mothers' work pathways and physical and mental health. *Journal of Health and Social Behavior, 53*(4), 396–412.

Friend, C. (2004). *Social contract theory*. www.iep.utm.edu/soc-cont/#H4. Accessed July 10, 2014.

Fuentes-Nieva, R., & Galasso, N. (2014). *Working for the few: Political capture and economic inequality*. Oxford, UK: Oxfam International.

Galli, A., Weinzettel, J., Cranston, G., & Ercin, E. (2013). A footprint family extended MRIO model to support Europe's transition to a one planet economy. *Science of the Total Environment, 461–462,* 813–818.

Galli, A., Wiedmann, T., Ercin, E., Knoblauch, D., Ewing, B., & Giljum, S. (2012). Integrating ecological, carbon and water footprint into a "Footprint Family" of indicators: Definition and role in tracking human pressure on the planet. *Ecological Indicators, 16,* 100–112.

GFN. (2014). *Footprint basics—Overview*. http://www.footprintnetwork.org/en/index.php/GFN/page/footprint_basics_overview/. Accessed August 6, 2014.

Gómez-Paredes, J., Yamasue, E., Okumura, H., & Ishihara, K. N. (2014). The labour footprint: An input–output based framework to assess labour dimensions of economic activity (submitted).

Hobbes, T. (1651a). LEVIATHAN. In C. B. Macpherson (Ed.). London: Penguin Books (1985).

ILO. (2012). *LABORSTA—Main statistics (annual): Employment general level, by economic activity, by occupation, by status in employment*. http://laborsta.ilo.org. Genève, Switzerland: International Labour Organization.

International Labour Organisation. (2014a). *Forced trafficking, human labour and slavery*. http://www.ilo.org/global/topics/forced-labour/lang–en/index.htm Accessed June 13, 2014.

International Labour Organisation. (2014b). *World of work report 2014: Developing with jobs*. Geneva: International Labour Office.

International Monetary Fund. (2007). *World Economic Outlook: Globalization and inequality*. Washington, DC, USA: International Monetary Fund, Publication Services.

Jackson, T., Jager, W., & Stagl, S. (2004) *Beyond insatiability; Needs theory, consumption and sustainability*. Working Paper Series Number 2004/2. Centre for Environmental Strategy, University of Surrey, UK.

King, I. (2014). The conflict over conflict-free minerals. *Accessed, 20*(07/14), 2014.

Kubiszewski, I., Costanza, R., Franco, C., Lawn, P., Talberth, J., Jackson, T., et al. (2013). Beyond GDP: Measuring and achieving global genuine progress. *Ecological Economics, 93,* 57–68.

Lenzen, M., Kanemoto, K., Moran, D., & Geschke, A. (2012b). Mapping the structure of the world economy. *Environmental Science & Technology*, *46*(15), 8374–8381. http://dx.doi.org/8310.1021/es300171x.

Lenzen, M., Moran, D., Kanemoto, K., & Geschke, A. (2012a). Building Eora: A global multi-region input-output database at high country and sector resolution. *Economic Systems Research* (submitted).

Lenzen, M., Moran, D., Kanemoto, K., & Geschke, A. (2013). Building Eora: A global multi-region input-output database at high country and sector resolution. *Economic Systems Research, 25*(1), 20–49.

Mason, R. (2014). Companies not asked to report slavery in supply chains under new laws. *The Guardian* [online], section.

McBain, D., & Alsamawi, A. (2014). Quantitative accounting for social economic indicators. *Natural Resources Forum* (n/a-n/a).

McBain, D., Wiedmann, T., & Lenzen, M. (2014). The contribution of input-output analysis to understanding impacts of consumption—A review update. *Journal of Industrial Ecology* (submitted).

McElroy, M. W., & van Engelen, J. M. L. (2012). *Corporate sustainability management: The art and science of managing non-financial performance*. Hoboken: Taylor and Francis.

Miller, R. E., & Blair, P. D. (2010). *Input-output analysis: Foundations and extensions*. Englewood Cliffs, NJ, USA: Prentice-Hall.

Minnaert, T. (2014). Footprint or fingerprint: international cultural policy as identity policy. *International Journal of Cultural Policy, 20*(1), 99–113.

Moran, D., McBain, D., Kanemoto, K., Lenzen, M., & Geschke, A. (2014). Global supply chains of Coltan: A hybrid LCA study using a social indicator. *Journal of Industrial Ecology*. Accepted May 2014.

Murray, J., & Wood, R. (Eds.). (2010). *The sustainability practicioner's guide to input-output analysis* (1st ed., 1 vols.). New York: Common Ground Publishing LLC.
Pelletier, N., Ustaoglu, E., Benoit, C., & Norris, G. (2013). Sustainability in trade and development policy. In Joint Research Centre, Institute for Environment and Sustainability (Eds.), *A life cycle approach to understanding and managing social risk attributable to production and consumption in the EU-27*. Luxembourg: Publications Office of the European Union European Commission.
Piketty, T. (2014). *Capital in the 21st century* (Trans.: A. Goldhammer). Cambridge, Massachusetts: Harvard University Press.
Prell, C., Feng, K., Sun, L., Geores, M., & Hubacek, K. (2014). The economic gains and environmental losses of US consumption: A world-systems and input-output approach. *Social Forces*.
Rees, W. E. (1992). Ecological footprints and appropriated carrying capacity: What urban economics leaves out. *Environment and Urbanization, 4*(2), 121–130.
Rees, W., & Wackernagel, M. (1996). Urban ecological footprints: Why cities cannot be sustainable—And why they are a key to sustainability. *Environmental Impact Assessment Review, 16*(4), 223–248.
Rose, A. (1995). Input-output economics and computable general equilibrium models. *Structural Change and Economic Dynamics, 6*(3), 295–304.
Rugani, B., Panasiuk, D., & Benetto, E. (2012). An input–output based framework to evaluate human labour in life cycle assessment. *The International Journal of Life Cycle Assessment, 17*(6), 795–812.
Solt, F. (2009). Standardizing the world income inequality database. *Social Science Quarterly, 90* (2), 231–242. SWIID Version 3.1, December 2011.
Tukker, A., & Dietzenbacher, E. (2013). Global multiregional input-output frameworks: An introduction and outlook. *Economic Systems Research, 25*(1), 1–19.
United Nations. (2012). *The future we want*. www.un.org/ga/search/view_doc.asp?symbol=A/RES/66/288andLang=E. Accessed August 4, 2014.
United Nations Environment Program. (2009). *Guidelines for social life cycle assessment of products*. Milan: United Nations.
United Nations Statistics Division. (2008). *The System of National Accounts (SNA)*. New York: United Nations.
United Nations Statistics Division. (2009). *System of National Accounts 2008*. New York.
UNSD. (2011). *National Accounts Official Data*. data.un.org/Browse.aspx?d=SNA. New York, USA: United Nations Statistics Division.
Ura, K., Alkire, S., Zangmo, T., & Wangdi, K. (2012). *A short guide to gross national happiness index*. Bhutan: The Centre for Bhutan Studies.
van den Bergh, J. C. J. M., & Grazi, F. (2014). Ecological footprint policy? Land use as an environmental indicator. *Journal of Industrial Ecology, 18*(1), 10–19.
Wackernagel, M., & Rees, W. (1996). *Our ecological footprint: Reducing human impact on the earth*. Gabriola Island, British Columbia, Canada: New Society Publishers.
Waring, M. (1988). *If women counted: A new feminist economics*. San Francisco: Harper Collins Publishers.
Weidema, B. P. (1993). Market aspects in product life cycle inventory methodology. *Journal of Cleaner Production, 1*(3–4), 161–166.
World Bank. (2014). Global economic prospects. In *Shifting priorities, building for the future*. http://www.worldbank.org/content/dam/Worldbank/GEP/GEP2014b/GEP2014b.pdf. Accessed August 6, 2014.
Zheng, H., & Land, K. C. (2012). Composition and decomposition in US gender-specific self-reported health disparities, 1984–2007. *Social Science Research, 41*(2), 477–488.

Chapter 7
Case Study—Assessing Social Impacts in the Seafood Industry

7.1 Introduction

Social impacts in global supply chains is a topic which has gone beyond corporate and civil society interactions for improved social (and environmental) outcomes, to a focus for consumers and the media. Over the past decade several raw materials/commodities have come to the forefront of discussions that go beyond technical and spill over into public ire and debate—palm oil, cocoa, conflict minerals and cotton to name but a few. Since 2014, Thai seafood has been one of those much-debated commodities.

There are several factors that often play into a commodity becoming a topic of focus. One is the country of origin. Although produced in a number of countries, the commodity is generally identified as having particularly negative impacts when from one particular country. In the case of palm oil, the strongest (and often most negative) association is with Indonesia, for cotton with Uzbekistan, and for seafood the association is with Thailand. This is not to say that these identified countries are the only countries with social issues related to these commodities. Looking at seafood, human trafficking in the seafood industry has been implicated throughout the industry from the Irish fishing industry[1] to Indonesia[2] to Taiwan.[3] And yet, current coverage of modern slavery issues would imply that the seafood industry in Thailand is the biggest concern. Retailers, consumers and brands have made decisions to not source from Thailand because it is seen as too high risk, based purely on the media profile. Is this the right thing to do? Does this improve the situation in Thailand? Is it justified given the implications of slavery in the seafood sector in so many countries throughout the world? This article considers the need

[1]https://www.theguardian.com/global-development/2015/nov/02/revealed-trafficked-migrant-workers-abused-in-irish-fishing-industry.
[2]http://www.ap.org/explore/seafood-from-slaves/over-300-slaves-rescued-from-Indonesia-island-after-ap-investigation.html.
[3]http://www.bbc.com/news/world-asia-27498048.

A. Alsamawi et al., *The Social Footprints of Global Trade*,
Environmental Footprints and Eco-design of Products and Processes,
DOI 10.1007/978-981-10-4137-2_7

for fair and equitable information on social impacts and supply chains, so that scientific and evidence based assessments of risk can be made.

7.2 Is Thailand the Highest Risk?

How can an assessment be made of a country/commodity risk profile? Many tools exist, but many are too blunt to fully assess risk or are not basing assessments on scientifically proven information. For example, a number of sustainability/social risk assessment tools will only identify problems to a country and commodity level, not to a specific supply chain level of detail. The Labor Safe Screen[4] and the Social Hotspot Database[5] are excellent examples of basing information on scientific, published data but that may not have the level of detail required for an in-depth analysis. The high level focus may be adequate for the first point of understanding that there may be an issue in a supply chain, but does not respect that different regions within a country can have significant variance on labor conditions such as migration, local regulations, growing conditions and markets.

In Thailand, migration is seen as one of the key risks of modern slavery, along with human trafficking and forced labour within the seafood industry. It is seen as one of the key risk factors for modern slavery. Much has been written in the media of migrants from Myanmar working in poor or even slave like conditions in the Thai seafood sector. However, migration from Myanmar is not consistent across the whole of the country—migrants travel from many countries surrounding Thailand to work there because the minimum wage is higher than in their own country, and there is an opportunity for work. Migrants will also tend to locate where their local communities are. For example, there is a concentration of immigrant workers from Myanmar in the Samut Sakhon province southwest of Bangkok in Thailand. This is due to a number of reasons, including but not limited to opportunities to work in the many factories and the port located here, but also due to the large Burmese population located in this province. Where you have a concentration of migrant workers, there are also benefits when it comes to addressing or resolving social impacts. Local regulators can focus their efforts to address critical issues, for example by adding additional departmental staff to process working visa applications. NGOs can focus their efforts in a concentrated area rather than dispersing across the whole country. This could include the provision of legal aid centres, shelters, grievance mechanisms and humanitarian support systems. You can also find that in a concentrated location of migrant workers or a particular industry sector, the industry itself can work together to address common

[4]http://www.laborsafescreen.com/.

[5]http://socialhotspot.org/.

problems through common solutions, for example through the provision of equal standards set and enforced through trade associations, or running open or common training programs. Migration in itself, whilst certainly being a risk factor for exploitation, should not be used as a proxy for modern slavery.

The US State Department publishes one of the most respected assessments of conditions of risk of modern slavery through their Trafficking in Persons (TIP) report. The 2016 TIP Report[6] identifies the Thai seafood sector as a high risk for modern slavery and human trafficking, and the subsequent human rights violations that go along with this (see p 363–368 for specific examples from Thailand of the seafood sector and other sectors such as the sex industry). This report is generated every year, and is the result of extensive research over each year. The report publishes progress, as well as areas that require more work. This year, the report upgraded Thailand from Tier 3 (the lowest grade) to the Tier 2 watch list in recognition of the work that had taken place between 2015 and 2016. Although this report is based on evidence (and not only media reports) it is still not without its controversy. In 2015, whilst evidence was still being found of human trafficking along a corridor between Myanmar, Thailand and Malaysia in the seafood sector (particularly of the Rohingya people) and implications of forced labour in the electronics and palm oil industries, Malaysia was controversially upgraded from a Tier 3 to a Tier 2. The implications were that this upgrade was due to political pressure rather than true advances in the treatment of people and human rights.[7,8] Therefore, even a government document, based on research, can be potentially misleading when trying to assess where social impacts are most likely to be occurring along a supply chain.

Media reports of risk can also be misleading. The Walk Free Foundation's *Global Slavery Index* report is another respected tool used to analyse country and sector risk for a wide range of factors including the production of commodities. The 2016 report found that an estimated 48.5 million people are in some form of modern slavery in 167 countries globally.[9] Of that figure, 58% live in five countries—India, China, Pakistan, Bangladesh and Uzbekistan. As an estimate of the proportion of the population in modern slavery, Thailand ranks at 20th in the world. Whilst certainly identifying a significant problem that needs to be addressed as a priority, many media reports give the Thai seafood sector a disproportionately large coverage. Articles such as *Thailand Ranked* 20th *on Global Slavery Index*[10] gave a much higher ranking to the Thai seafood sector than it might otherwise deserve on a risk basis, and many online publications of articles (often with a fair depiction of the report) lead pictures

---

[6]http://www.state.gov/j/tip/rls/tiprpt/2016/index.htm.

[7]https://www.theguardian.com/global-development/2015/jul/27/us-human-trafficking-in-persons-report-under-fire-cuba-malaysia-upgraded.

[8]http://www.reuters.com/article/us-usa-humantrafficking-disputes-special-idUSKCN0Q821Y20150804.

[9]http://www.globalslaveryindex.org/findings/.

[10]https://www.undercurrentnews.com/2016/06/02/thailand-ranked-20th-on-global-slavery-index/.

of fishing vessels and fish,[11,12,13]. If we are to truly understand the social impact risks in global supply chains, we need to move beyond sensationalism and address the current reality, recognising critical problems as well as critical solutions. Sadly, far fewer awards are being won for writing about solutions to modern slavery than identifying the problems.

7.3 Key Risk Factors Around Modern Slavery and Human Rights in the Seafood Sector

Although difficult to identify the key issues and risk factors in modern slavery through reports or the media, insight can be gained through talking to people working in the industry in the location of concern, and by engaging with civil society working on those issues based locally. As discussed previously, there are many migrants working in the seafood sector (in Thailand and around the world) and this escalates risk. One of the key contributors to the risk is debt bondage, where workers build up a debt either getting to their destination country or while working there, often so great that they cannot hope to repay it. Debt bondage can arise in the seafood sector during the migration process, through brokers charging jobseekers to find work, through brokers charging a percentage of salary for processing of wages as well as charging for accommodation, food, transport, uniforms or even transport to work. Workers are not always aware of their rights or benefits under local laws, and therefore less likely to pursue fair treatment or justice. Even if workers are aware of their rights, facilities or associations to pursue fair treatment often do not exist. In these situations, allowing freedom of association is important, so that workers can jointly express their need for fair and equal treatment and processes. Correct and legal documentation is often an issue, particularly where the conditions for safe and legal migration between governments of neighbouring countries is not clear and borders are relatively porous. Withholding of key documents such as passports and work permits remains an issue, which when teamed with unpaid wages makes it difficult if not impossible for workers to escape from modern slavery.

An overarching concern within the seafood sector, is labour at sea. Unlike the manufacturing and agricultural sectors, where facilities and farms cannot move, fishing vessels are mobile and can easily and quickly move out of range of land, and even out of range of a country's territorial waters and into international waters. Factory auditing processes that have been in place for many years, to standards such

---

[11]http://www.bbc.com/news/world-asia-36416751.

[12]https://www.theguardian.com/global-development/2016/jun/30/slavery-exploitation-worst-offender-lists-biased-towards-rich-countries-wont-help.

[13]http://www.abc.net.au/radionational/programs/drive/global-slavery-index-estimates-there-are-45.8-million-people-in/7472416.

as SA800, BSCI or SEDEX's SMETA code have not taken place on vessels because the workplaces are frequently not comparable. At sea there is no easy way to monitor work hours and breaks, labour conditions or even give workers a voice to raise an alarm if there is a serious risk to health or life. Conditions on vessels are, by their nature, confined, and the work is generally physically demanding. This is not to say that great improvements in the sector are not being made. There are now audits of vessels at sea as well as factories on land by government authorities, NGOs and by concerned industry players. Leaps in technology are bringing traceability to sea and with it worker access to voice, help and remedy. The global focus on the issues, whether through local or international government pressure, consumer and NGO pressure, tighter legislative requirements or industry leadership, is bringing greater transparency and a faster pace of change.

7.4 The Unanswered Question—Who Pays for Improved Working Conditions?

One of the biggest issues, well beyond just the Thai seafood sector, is increasing world demand and consumption, for reduced prices. Many retailers pride themselves on the provision of quality goods at the lowest possible cost. There may be some efficiencies to be gained along global supply chains, but if a supply chain is under serious pressure to deliver lower costs or lower margins, then wages and work conditions (not to mention environmental conditions) upstream in the supply chain are often the first to suffer. It is not logical to expect purchase price to decrease if at the same time there is an expectation for an increase in salaries, training, occupational health and safety and reduced working hours.

Consumer research shows that, when asked, people would like to buy products that are produced ethically and workers are paid a fair wage. Whilst research shows that people are willing in theory to pay more for a sustainable certified brand if there is a logo that they recognize, this does not always translate into practice. Recent research by the Marine Stewardship Council (MSC)[14] conducted by GlobeScan showed that in some age brackets up to 75% of respondents believed that eating sustainably sourced seafood (with respect to ocean sustainability, not specifically labour conditions) was important and over half (54%) of respondents said that they were willing to pay more for a certified sustainable seafood product. However, only about 10% of the world's wild caught seafood comes from MSC certified fisheries (and not all of that is labelled as MSC certified) and the standard does not cover any social impacts such as wages, forced labour or human trafficking. Currently no label to this standard exists for seafood, and therefore it is difficult to provide a mechanism that consumers will trust to pass money from point of sale back to the

[14]https://www.msc.org/documents/msc-brochures/msc-consumer-survey-2016-infographic-seafood-consumers-put-sustainability-before-price-and-brand.

workers upstream in the supply chain. The issues and reasons for human trafficking and forced labour are so complex and multifaceted that no one company or organization is likely to be able to resolve the issues themselves. Ultimately, discussions and commitment to pay a fair wage all along a supply chain requires involvement and financial commitment from all of the commercial participants, as well as genuine engagement with civil society and NGOs.

7.5 Conclusion

The Thai seafood sector is complex, and the pressures on the industry to reform are great. Excellent examples of progress can be found,[15,16,17]although perhaps these examples are still too few. In analyzing social impacts in global supply chains, the Thai seafood sector teaches us that we need to rely not only on statistics, but qualitative research and amplifying good practice rather than vilifying bad to the exclusion of all else. As for many commodities, where an issue arises in one country you are more likely to find it in others than to find an isolated example. Ultimately, supply chains solutions need to be found as just that—involving consumers, civil society, retailers, brands, manufactures and producers in the same conversation about priorities and what price we put on equality.

[15]http://www.projectissara.org/.

[16]http://www.thaiunion-sustainability.com/.

[17]http://www.shrimptaskforce.global/.

Chapter 8
Raising the International Poverty Line—A Comparison of Necessary Adjustments of Final Demand Spending in OECD and Non-OECD Countries

8.1 Introduction

In today's globalized world, goods and services consumed in one country require inputs from a great variety of other countries. These inputs not only include raw materials and machinery but also labour. In many countries, workers, especially in the primary sector, are not paid an income high enough to provide a decent living standard to them and their families. Thus, despite having a job they may live in extreme poverty. Target 1.1 of the Sustainable Development Goals, that were adopted be the UN General Assembly in September 2015 (UN 2015), specifies that by 2030 extreme poverty, measured as people living on less than \$1.25 a day (in purchasing power parities), should be eradicated for all people everywhere. In December 2015, the World Bank announced that everyone living on less than \$1.90 a day is considered to live in extreme poverty, thus raising the poverty line by \$0.65 (Ferreira et al. 2015). Note that these numbers are given international purchasing power parity dollars, thus reflecting the living costs in different countries.

Multi-regional input-output systems can be used to link the consumers of goods and services to those who are producing the intermediate goods and services required as inputs into the final products. Using the EORA multi-regional input-output system (Lenzen et al. 2013) combined with data on the number of workers living below the poverty line, we show where in the world products manufactured by workers living below the poverty line are consumed and how this compares to products manufactured by other workers (those not living below the poverty line). To eradicate extreme poverty, the workers living below the poverty line of \$1.25 would need to earn significantly more. If consumers liked to continue consuming the same products (goods and services) and the same amount of products, they would need to pay more for this. Using MRIO analysis, it is possible to estimate necessary final demand spending if the income of the workers living below the poverty line was to be raised from the old poverty line of \$1.25 to the new poverty line of \$1.9 (this may even be more than an increase by \$0.65 as not all

A. Alsamawi et al., *The Social Footprints of Global Trade*,
Environmental Footprints and Eco-design of Products and Processes,
DOI 10.1007/978-981-10-4137-2_8

workers living below $1.25 actually earn $1.25 a day, but maybe significantly less). In the discussion below we differentiate between OECD and non-OECD countries. The 35 countries who are members of the OECD (Organisation for Economic Cooperation and Development) are among the world's most advanced countries[1] with high per capita income and correspondingly high per capita final demand.

The chapter is organized as follows: the next section presents the data underlying the poverty footprint calculations. Section 8.3 describes the poverty footprint of OECD and non-OECD countries and Sect. 8.4 shows the results of the policy simulation of raising the workers' income to the old and new poverty lines. Section 8.5 concludes.

8.2 Population, Employment, Working Below the Poverty Line and Final Demand

In 2010, 18% of the world population lived in the countries that are currently member of the OECD. These countries comprised 21% of the global labour force, but less than 5% of the workers living below the poverty line. The EORA MRIO database contains 187 countries and includes the economic interactions between industries within countries, between industries in different countries and the interaction of industries and final consumers across the world. There are a variety of extensions to the MRIO, i.a. CO_2 emissions, biodiversity, employment and poverty indicators. In our analysis below we are focussing on the employment (Alsamawi et al. 2014) and poverty (Alsamawi et al. 2016) extensions of EORA for 2010. Unfortunately, the poverty data covers only 121 countries, with those countries missing that have especially high prevalence of poverty.[2] Data on OECD countries are complete. The percentages below are reported from the total numbers available, thus excluding the missing countries. Thus, due to missing data, some percentage numbers for OECD countries are significantly overestimated and accordingly, some percentage numbers for non-OECD countries are significantly underestimated. Figure 8.1 illustrates the shares of OECD and non-OECD countries in workers below the poverty line, employment, population and final demand.[3] At least 95% of the world's people working below the poverty line are in non-OECD countries,[4] while the shares of people employed in these countries and their population are only at least 75%. The OECD's share in workers below the poverty line is dominated by

[1]http://www.oecd.org/about/membersandpartners/.

[2]More than half of the Sub-Saharan African countries and some of the Western Asian countries including India are missing as well as some of the Latin American countries and Pacific Island states.

[3]Note that the shares relate to those countries for which poverty data (i.e. data on the number of workers with daily earnings below the poverty line) is available.

[4]Note that the poverty line of $1.25 is calculated in international purchasing power parities. That is, in more expensive countries, the actual dollar value of the poverty line is significantly higher.

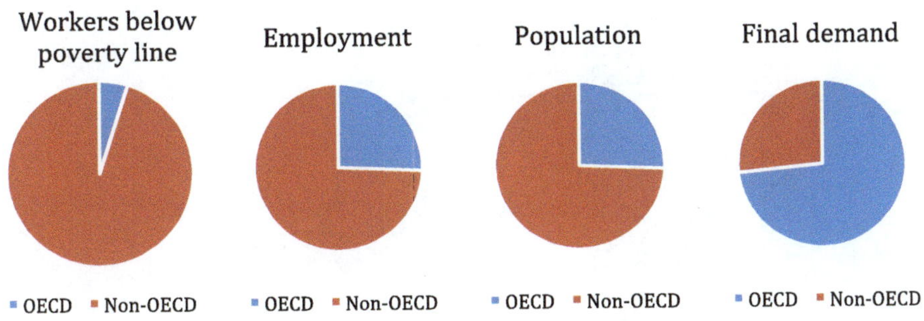

Fig. 8.1 Shares of OECD and non-OECD countries

three countries: more than 6 million workers in Mexico, 2.7 million workers in the USA and 1.2 million workers in Turkey.

Global final demand is dominated by OECD countries (almost three quarters). Table 8.1 gives an overview on all the underlying data used here and its availability. The data shown here is the "production perspective", i.e. the workers are allocated to the countries where they are part of the production of intermediate and final goods. This is in contrast to the "final demand" or "footprint" perspective discussed in the next section.

8.3 "Working Below the Poverty Line" Footprint

The term footprint has been introduced in the ecological analysis literature by (Rees 1992) as "ecological footprint" and has been widely used since, see e.g. (Wiedmann et al. 2006; Turner et al. 2007), and it refers to environmental impacts of upstream production processes. The ecological impacts are allocated to those countries where the final products are eventually consumed (used to satisfy final demand). Here, the footprint terminology is transferred to "working below the poverty line" footprint indicating the number of workers involved in upstream production processes earning less than the poverty line. "Working below the poverty line" footprint will be referred to as poverty footprint for the remainder of this chapter. The poverty footprint is thus the number of workers earning below the poverty line, who are engaged in the production chain of goods and services needed to satisfy final demand in a country.

To calculate the number of workers (total employment) or workers below the poverty line engaged in producing intermediate and final goods to satisfy in final demand, i.e. the employment and the poverty footprints, the usual MRIO methodology is applied. For this, define the stressors S_i^c as the total number of workers engaged in the production of one unit of output ($S_i^c = E_i^c$) or the number of workers below the poverty line engaged in the production of one unit of output ($S_i^c = P_i^c$) in industry i in country c (with the number of industries in country c denoted as n_c and

Table 8.1 Data availability and summary statistics

	Workers below the poverty line			Total employment			Population			Final demand (in USD pp)		
	OECD	Non-OECD	Total	OECD	Non-OECD	Total	OECD	Non-OECD	Total	OECD	Non-OECD	Total
n	35	86	121	35	106	141	35	149	184	35	152	187
Total (in mio)	13.1	251.1	264.2	551.1	2120.2	2671.3	1243.6	5637.4	6881.0	43,323	19,110	62,432
Total (in mio) with available poverty data (n non OECD 86)	13.1	251.1	264.2	551.1	1599.9	2151.0	1243.6	3607.8	4851.4	43,323	15,710	59,033
Share (%)	5	95	100	21	79	100	18	82	100	69	31	100
Share in population (%)	1	4	4	44	38	39						
Share in population (available countries) (%)	1	7	5	44	44	44						
Share in total employment (%)	2	12	10									
Share in total employment (available countries) (%)	2	16	12									

the total number of countries denoted as N). $(\mathbf{I}-\mathbf{A})^{-1}$ is the global Leontief inverse for all industries in all countries and $\mathbf{Y}$ the corresponding global final demand matrix. The stressor footprint $\mathbf{sf} = colsum(\mathbf{SF})$, that is employment footprint $\mathbf{ef} = colsum(\mathbf{EF})$ and poverty footprint $\mathbf{pf} = colsum(\mathbf{PF})$, is calculated from

$$\mathbf{SF} = \mathbf{S}^{row}(\mathbf{I}-\mathbf{A})^{-1}\mathbf{Y}$$

$$\text{with } \mathbf{S}^{row} = \begin{bmatrix} S_1^1 \ldots S_{n_1}^1 & 0\ldots0 & 0\ldots0 \\ 0\ldots0 & S_1^2\ldots S_{n_2}^2 & 0\ldots0 \\ & & \\ 0\ldots0 & 0\ldots0 & S_1^N\ldots S_{n_N}^N \end{bmatrix},$$

so that the size of $\mathbf{SF}$ is $(N \times N)$, the rows reflecting the countries where the workers are employed and the columns reflecting the countries where the final products are consumed.

Figure 8.2 shows the "poverty footprint", that is the number of workers employed below the poverty line to satisfy the final demand of a given country. The countries with the highest poverty footprint are Brazil, Bangladesh, Thailand, Pakistan and Indonesia, all non-OECD countries with a large population. The OECD countries with the highest poverty footprint are the USA, Japan, Germany, Mexico and Switzerland.

Overall, 28% of the global poverty footprint can be linked to final demand in OECD countries, while the remaining 72% are linked to final demand in non-OECD countries. This is due to the fact that a substantial share of domestic production is still aimed at domestic final consumption, so that the domestic effect dominates. Nonetheless, international trade has a significant impact on the

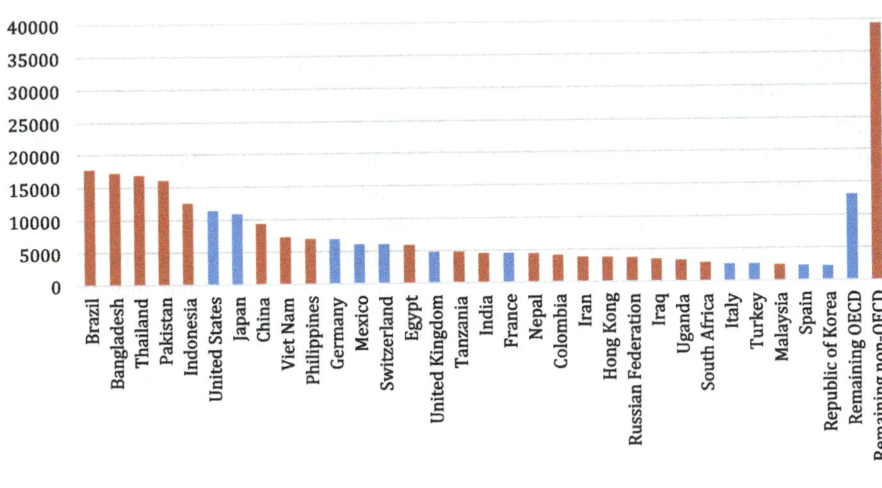

Fig. 8.2 Workers below poverty line to satisfy final demand (in thousands)

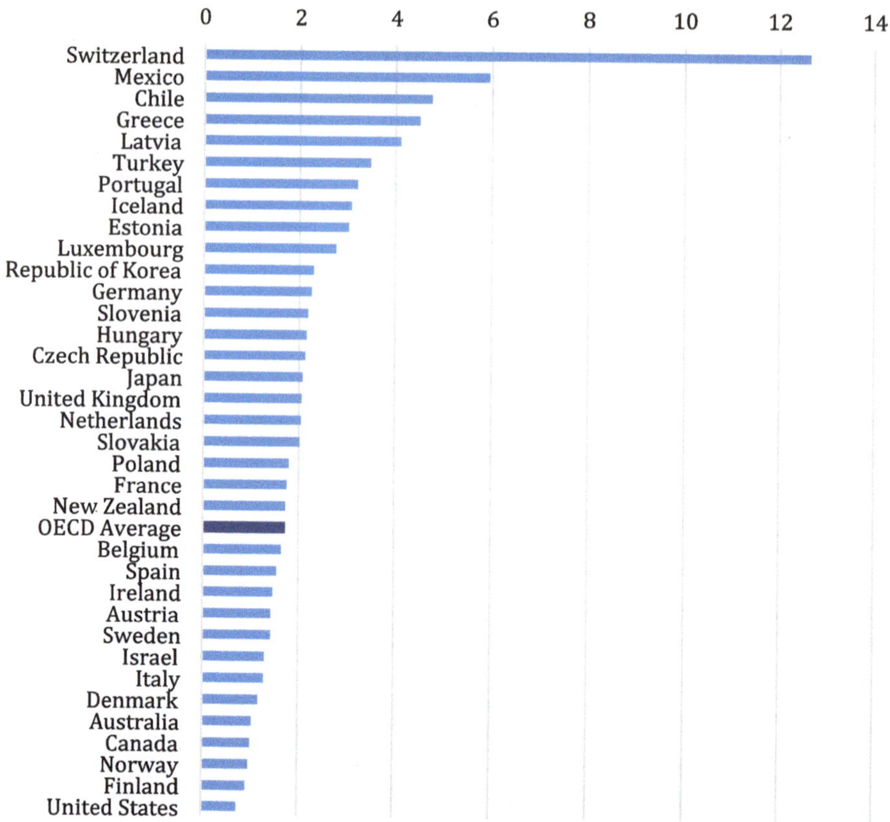

Fig. 8.3 Workers below poverty line to satisfy final demand per million USD of final demand

allocation of workers earning below the poverty line, when looking at the final demand perspective in comparison to the production perspective.

For Brazil 91% of the poverty footprint is domestic, i.e. 91% of the workers earning below the poverty line, who are engaged in production for Brazilian final demand, are employed in Brazil. The 99.5% domestic share in Bangladesh is even higher. But here, the missing poverty data for India might play a significant role as India is the second highest import origin[5] of Bangladesh. Thus, if poverty data for India were available the foreign share of Bangladesh's poverty footprint would increase. Thailand's domestic share is 93%, Pakistan's is 98% and Indonesia's is 94%. All of these countries are in close proximity to India as well, thus their actual domestic is likely to be lower.

For most OECD countries, the domestic shares are negligible. For these countries it is more interesting to look at the "poverty intensity of final demand", i.e. how many workers are employed below the poverty line to satisfy their final demand. Figure 8.3

[5]http://atlas.media.mit.edu/en/profile/country/bgd/.

shows the number of workers earning below the poverty line per million USD final demand. The high number for Switzerland is mainly driven by workers in Russia, while for Mexico and Chile the intensities are driven by the domestic shares (85 and 70%). The OECD average it 1.7 workers below the poverty line per million USD final demand, with most countries being between 1 and 2.3. As mentioned above, the poverty data is missing for a substantial number of countries, so that the numbers here are only lower bounds and could easily be twice as high or even higher.

8.4 Policy Scenarios: Raising Workers' Wages to Poverty Line

To calculate the necessary additional final demand spending to raise all workers' wages to the poverty line, it is necessary to make an assumption about the current average wage of workers below the poverty line. Here, 0.25 PPP$ are chosen, i.e. for the workers to earn as much as the old poverty line 1 addition PPP$ per worker is necessary. To increase their wage to the new poverty line 1.65 PPP$ per worker are necessary. These numbers need to be multiplied by 365 days of the year to get the annual increase.

The results show that the OECD countries on average would need to spend 0.062% more on final demand to lift workers in their poverty footprint from below the poverty line, to old poverty line of 1.25 PPP$. This is far below the international target for official development aid of 0.7% of annual GDP. An additional 0.04% of final demand spending is necessary to lift workers up to the new poverty line of 1.90 PPP$. For detailed country results, please see Fig. 8.4. Those countries with High poverty footprints per dollar of final demand also need to increase their final

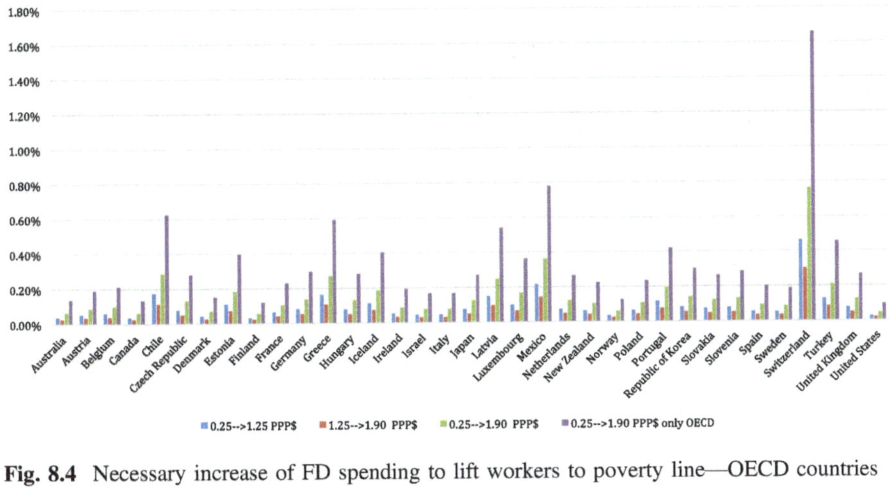

Fig. 8.4 Necessary increase of FD spending to lift workers to poverty line—OECD countries

demand spending the most: Switzerland 1.8% (for getting from 0.25 PPP$ to 1.9 PPP$), Mexico 0.36% and Chile 0.29%.

The non-OECD countries would need to spend 0.36% more on final demand on average to get to 1.25 PPP$ wages of workers in their poverty footprint and an additional 0.24% to get workers' earnings up from 1.25 PPP$ to 1.9 PPP$. As these increases in final demand spending is achieved through price increases of the final products, the purchasing power of the people in non-OECD countries would be reduced more than the purchasing power of people in OECD countries. Thus, if all the burden of the increased wages would be borne by OECD countries, additional final demand spending in the OECD countries would on average need to be 0.22% for the poverty line of 1.25 PPP$ and 0.36% for the new poverty line of 1.9 PPP$.

One big caveat in this analysis is that the poverty data only covers 86 out of 152 non-OECD countries and that these data only cover the workers, not their families or other people living below the poverty line. Thus, the numbers presented above are only lower bounds on necessary additional final demand spending.

A quick estimation using data on the world population for 2010 (6.924 billion people),[6] and the percentage of the world's population living below the poverty line (16.3%),[7] gives an estimated 1.129 billion people living below the poverty line. The poverty data presented here include 264.2 million workers (see Table 8.1), thus only about one quarter (23.4%) of the world's population living below the poverty line. Hence to lift all of these people to the poverty line of 1.9 PPP$, the OECD countries would need to spend 0.36% * 1/(23.4%) = 1.54% on final demand. This is about twice as much as the agreed target of 0.7% of GDP for official development aid. However, the OECD countries are not the only high income countries and the burden per country could be lowered of other high income countries would take some share of the responsibility as well. Btu even if not, 1.5% of additional spending could be worthwhile for a more equal world.

8.5 Discussion

This raise in wages could be viewed as setting a global minimum wage. Such a global policy is difficult to implement, but many countries already have set national minimum wages (WageIndicator Foundation 2016). For Bangladesh for example, the minimum wage is Bangladeshi Taka 1500 per month which is less than US$20 at current exchange rates and thus significantly less than the poverty line [international PPP dollars which are used for measuring the poverty line are pegged to the US$ with a factor of 1 (World Bank 2016)]. Raising the minimum wage and, with that, increasing the price for final consumers, maybe only in High Income Countries (not only in OECD countries), would be a very different approach to

[6]http://data.worldbank.org/indicator/SP.POP.TOTL.

[7]http://data.worldbank.org/topic/poverty?end=2012&start=1981.

tackle the global poverty reduction problem. (Bourguignon et al. 2002) review different economic methodologies to analyse poverty reduction policies. The policies they cover are broad, ranging from cash transfers and public spending to tax policies, but nothing like setting a minimum wage. One of their findings is, that the analysis should be done at the marginal level, not using indicators such as average income. The global minimum wage policy would be such a policy and using MRIO analysis the economic effects of this poverty reduction policy can be analysed on the meso (industry) and macro (country) level. The analysis on the industry level needs to be done from both, production (workers) side and consumption (final demand) side, if possible even differentiating between final demand of different income groups as has for example been done by (Hubacek Co-et al. 2015). Then also distributional effects on both ends can be considered.

References

Alsamawi, A., Murray, J., & Lenzen, M. (2014). The employment footprints of nations. *Journal of Industrial Ecology, 18*(1), pp. 59–70. Available at: http://doi.wiley.com/10.1111/jiec.12104. Accessed June 7, 2016.

Alsamawi, A., Murray, J. & Lenzen, M. (2016). Poverty and trade: Poverty footprints of global supply chains. *Unpublished*, pp. 1–30.

Bourguignon, F., et al. (2002). Evaluating the poverty impact of economic policies : Some analytical challenges.

Ferreira, F. H. G., et al. (2015). A global count of the extreme poor in 2012 data issues, methodology and initial results.

Hubacek Co-, K., et al. (2015). Global income inequality and carbon footprints topic: Tracing carbon emissions in global value chains I. In *23rd Conference of the International Input-Output Association*. Mexico City. Available at: https://www.iioa.org/conferences/23rd/papers/files/2184.pdf. Accessed July 26, 2016.

Lenzen, M., et al. (2013). Building Eora: A global multi-region input-output database at high country and sector resolution. *Economic Systems Research*, 25(1), pp. 20–49. Available at: http://www.tandfonline.com/doi/abs/10.1080/09535314.2013.769938. Accessed April 8, 2016.

Rees, W. E. (1992). Ecological footprints and appropriated carrying capacity: What urban economics leaves out. *Environment and Urbanization, 4*(2), pp. 121–130. Available at: http://eau.sagepub.com/cgi/doi/10.1177/095624789200400212. Accessed July 26, 2016.

Turner, K., et al. (2007). Examining the global environmental impact of regional consumption activities—Part 1: A technical note on combining input–output and ecological footprint analysis. *Ecological Economics, 62*(1), pp. 37–44. Available at: http://linkinghub.elsevier.com/retrieve/pii/S0921800906005933. Accessed November 29, 2012.

UN. (2015). *Transforming our world: The 2030 agenda for sustainable development.*

WageIndicator Foundation. (2016). Minimum Wages. *Powered by University of Amsterdam-AIAS. FNV. Career site Monster. WageIndicator Foundation—Share and compare wages, understand Labour Laws and spot career opportunities.* Available at: http://www.wageindicator.org/main/salary/minimum-wage. Accessed July 26, 2016.

Wiedmann, T., et al. (2006). Allocating ecological footprints to final consumption categories with input–output analysis. *Ecological Economics, 56*(1), pp. 28–48. Available at: http://linkinghub.elsevier.com/retrieve/pii/S0921800905002600. Accessed October 27, 2012.

World Bank. (2016). Table 4.16: Exchange rates and prices. *World development indicators.* Available at: http://wdi.worldbank.org/table/4.16. Accessed July 26, 2016.

Chapter 9
The Inequality Footprints of Nations; A Novel Approach to Quantitative Accounting of Income Inequality

9.1 Introduction

There is no doubt that inequality in income exists both within countries and between countries. There also seems to be agreement that in both cases it is rising (UN 2009; UN System Task Team 2012). The causes of this inequality are attributed variously to: conflict, governance and possession of natural resources (Collier 2007); technological change (OECD 2011a); jobs; transportation costs (Krugman and Venables 1995); and globalisation (Pavcnik 2011; Goldberg and Pavcnik 2007).

The UN sees inequality as a social justice issue as well as a threat to social, political and economic stability around the world (UN Department of Economic and Social Affairs 2010; UN 2005). Recent studies also link inequality to environmental problems, sustainability, crime, disease and overall well-being which in turn is closely linked to life expectancy (HM Government 2005; UN Systems Task Team 2012; Jackson 2012).

Attributing to inequality such dire global consequences as social, political and economic instability has implications for global action. Even though there are no simple solutions to what has been an intractable problem throughout history, our recognition of inequality and its consequences brings with it obligations. Competition for the consumer dollar has driven demand for ever-cheaper goods and services. Our demand sets in motion supply chains that ripple around the world. From time to time scandals erupt in the press, such as use of child labour in the making of a specific product (Burke 2000; Foley 2009). In such cases we take some responsibility for our consumption and pressure global businesses to improve their practice in third-world countries (Basu and Zarghamee 2009). However, in the case

Alsamawi, A., Murray, J., Lenzen, M., Moran, D., Kanemoto, K. (2014). The inequality footprints of nations: A novel approach to quantitative accounting of income inequality. *PLOS ONE, 9*(10): e110881. doi:10.1371/journal.pone.0110881.

A. Alsamawi et al., *The Social Footprints of Global Trade*,
Environmental Footprints and Eco-design of Products and Processes,
DOI 10.1007/978-981-10-4137-2_9

of inequality existing within a particular country the relationship to exported goods is not so clear-cut. In consuming goods from this country—any goods—are we implicated in the inequality that exists in the exporting country? If yes, the simple solution would be to pressure the country to change through boycotting its goods just as we boycott firms for their use of child labour. However without our demand there would be no production and perhaps, as cited in the case of child labour, no jobs for those desperately in need of work (Basu and Zarghamee 2009). Importing from unequitable countries could actually ameliorate the situation. In this study we ask simply: do more egalitarian countries import from less egalitarian ones? Dissecting the social impacts of a marginal dollar of trade with a country suffering from inequality is a difficult task. But regardless of whether trade between equal and less equal countries leads to a net gain or loss of welfare, understanding where unequality is occurring, who is benefiting from it, and which countries have the most polarized trade, provides information useful for understanding the dynamics of inequality and trade.

This study introduces the novel concept of an 'inequality footprint', which is defined as the Gini index of the workforce that is directly and indirectly required to satisfy the consumption of a given population. Thus, the inequality footprint extends beyond the boundaries of a particular country and includes people working in countries that produce goods and services bound for international trade. In this work we undertake a quantitative analysis of the inequality footprint of the world's nations and portray these footprints using a number of intuitive measures. We do not argue whether trade between more and less equal countries is a social good or ill, but merely provide a robust, global, account of inequality and trade upon which such further economic analysis may be built.

In Sect. 9.2 we provide a context for our work through an overview of the inequality literature, discussing the issue from the perspective of world bodies such as the UN and the IMF and drawing on some of the within-country inequality literature. Following that we provide our methodology and our data sources in Sect. 9.3. We present the results in Sects. 9.4 and 9.5 concludes.

9.2 Background

9.2.1 Inequality as a Global Issue

The first target of the first millennium development goal (MDG1) is to halve, between 1990 and 2015, the proportion of people whose income is less than $1 a day.[1] Growing inequality makes it harder to reach this goal. As Kofi Annan observed in 2005: "we cannot advance the development agenda without addressing the challenges of inequality within and between countries" (UN 2005, p. iii). Over

[1]http://www.un.org/millenniumgoals/poverty.shtml (accessed 01/10/12).

the past 20 years inequality has risen within more countries than it has fallen (United Nations Development Program 2010). In 2007 the United Nations Assistant Secretary-General for Economic Development, Jomo Kwame Sundaram, talked about a significant and disturbing increase in inequality within and between countries around the world,[2] which he attributed to a worldwide decline in the number of jobs. Even in countries in Northern Europe that have the world's lowest income inequalities (International Monetary Fund 2007), such as Germany, Denmark and Sweden (OECD 2011a) inequality grew in the 2000s probably because of increased income disparity (Dreher and Gaston 2008).

Atkinson et al. (2011) see increasing income disparity as at least partly attributable to an unprecedented surge in top wage incomes (see also OECD 2011b). They attribute the evolution of the top one per cent share of wealth in various countries to political changes (e.g. Reagan in the US and Thatcher in the UK), wars, financial crises, global factors and taxation. Whatever the reason the wealthiest 20% now account for 86% of all private consumption and the poorest account for around one per cent (UN Department of Economic and Social Affairs 2010). Even in China and Russia increases in income are invariably going to the top 1% of the population (Atkinson et al. 2011; Denisova 2012). This matters because inequality within countries affects people's well-being (UN System Task Team 2012; Jackson 2012); it breeds social resentment and generates political instability as people feel that they are losing out while others are becoming rich (OECD 2011a; Chapple et al. 2009). Such inequality would seem to be a recipe for disaster.

9.2.2 *Within Country Inequality*

In the past it was assumed that globalisation would raise the income of almost all nations and help redress inequality (Krugman and Venables 1995). However studies have found no obvious relationship between a country's participation in globalisation and changes in inequality within countries (International Monetary Fund 2007; Dreher and Gaston 2008). A 2008 study of wage inequality in Mexico, for example, found increased access to export markets led to growing wage inequality[3] (Verhoogen 2008) with the need to produce better products for export than for domestic markets within industries and hence the need for upgrading of skills for some workers. Thus wage disparity increased within manufacturing industries where smaller less productive plants with less skilled labour supplied local markets while larger more productive plants with highly skilled labour

---

[2]UN News Centre http://www.un.org/apps/news/story.asp?NewsID=21526&Cr=globalization&Cr1#.UF9-vRiTbbs (accessed 24/09/12).

[3]Although a study conducted by the IMF (2007: 35) shows inequality in Mexico fell after the mid-1990s (see also Goldberg and Pavcnik (2007). Distributional Effects of Globalization in Developing Countries. *Journal of Economic Literature, 45*(1), 207–232.).

produced goods for export. Other studies quoted by Pavcnik (2011) found a similar process in Argentina and Indonesia (see also UN 2005; Denisova 2012).

A related skill-based driver of wage inequality has been a general increase in use of information and communication technologies (ICT) by some industries, which has increased their demand for high-skilled, college educated, workers (Pavcnik 2011; International Monetary Fund 2007). Where access to technology skills and education are available only to the elite the poor have no ability to move from unskilled to higher skilled occupations (e.g. from agriculture to industry) thus increasing inequality. This selective apportioning of education is not confined to developing countries. A 2005 UN study found some of the highest income inequalities—as well as in Mexico and Turkey—were in the USA (UN 2005) much of which in the US[4] was attributed to access to education, in particular skill-biased technological change (SBTC) (Galor and Moav 2000; Borjas and Ramey 1995). Equal access to education, job creation and skill development opportunities are consistently seen to be the most important factors in the building of a more equitable society (OECD 2011a; Pavcnik 2011).

The issue around access to education and its effect on inequality is not confined to the production of finished goods. Goldberg et al. (2010) suggest that global trade in intermediate goods also has an effect on wage inequality with some large firms from developed countries locating unskilled parts of production in developing countries and reserving any skilled part of production for developed countries (Pavcnik 2011). Apart from relieving firms of the need to develop skills in workers in developing countries, this can also have the effect of firms shifting responsibility for wages and working conditions away from the regimes of developed countries where conditions are generally governed by industrial law to countries where such laws do not exist or else are only just emerging. In this case (i.e. production of intermediate goods) and in the case of final products the producer has some control over skill development, working conditions and wages and can act to alleviate inequality if motivated to do so.

At the other end of the supply chain consumers have some control over the demand for imported goods. For example consumers in Scandinavia live in some of the most equal of all OECD countries (OECD 2011b) and amongst the most equal in the world (International Monetary Fund 2007) yet they import goods from some of the most unequal. One such is China where although foreign demand has increased employment considerably (International Monetary Fund 2007) it has induced jobs mainly in low-paid, low and medium-skilled areas without generating new job opportunities for the growing number of college-educated workers in the middle thus adding to inequality). Another Scandinavian trading partner, the Russian Federation, also has high inequality (UN 2010). The economic recessions in Asia, Latin America and the Russian Federation following the financial crises of

[4]Coincidentally the life expectancy of poorly educated women in the US has also slipped in the last decade from a gap of 5.8 years to a gap of 8.6 years compared with women with a Bachelor's degree or higher (Norton and Ariely (2011). Building a Better America—One Wealth Quintile at a Time. *Perspectives on Psychological Science, 6* (1), 9–12).

the 1990s brought increased unemployment and inequality, both of which contributed to the erosion of social cohesion. This was especially true of Russia, where the Gini inequality index rose from 0.397 in 2001 to 0.422 in 2009 (Denisova 2012) and which recorded one of the lowest life satisfaction scores in the world just above Moldova, Ukraine and Belarus (Jackson 2012). Other Scandinavian trading partners such as Brazil and the Philippines are experiencing worsening inequality, which has been linked to trade liberalisation. For example a 2005 UN study (UN 2010) found that trade liberalisation caused wages to decline in Brazil and Mexico especially in the case of unskilled workers. The study went on to also link trade liberalisation with widening inequality in the Philippines and Eastern Europe.

9.2.3 Responsibility

The debate around the causes of within-country inequality, and even within which countries it is rising or falling, is unresolved (Basu 2005). Also different methodologies for studying inequality can produce different results (Milanovic 2005) making it difficult to generalise causes and specific effects. So who may be held responsible for improving the situation? Blaming bad government in such countries (Collier 2007) is to oversimplify the issues. After all, global companies do business with these countries. As Goldberg et al. (2010) point out, increasingly global trade is about movement of goods between firms located in different countries rather than the flow of goods between countries, so perhaps such firms should take the blame for supporting—and possibly exploiting—inequality. Blaming the final consumer for demanding goods contaminated by the implication that they arose from countries of gross inequality is also a simplification. Consumers could also be seen as providing jobs for the desperately poor. People need to work and often a poorly paid job in a country of high wage disparity is better than no job. Firms, governments and consumers all play a part in the complex web of production, wages and work. With a better understanding of within-country inequality all can play a part in bringing about a more equal society.

The ‘inequality footprint’ can provide a tool to assist in tracking the inequality implicated in goods as they move around the world. The concept of ‘footprint’ is familiar from the literature. Studies have been conducted, for example, on the emissions footprint showing that consumption in one country can cause emissions in another (Wiedmann et al. 2008). Or the water footprint that can track embodied water in goods as they move around the world and can distinguish ‘scarce water’ inputs from non-scarce water inputs (Hoekstra and Chapagain 2007). We also know that consumption in one country requires the input of labour from other countries (Los et al. 2012), which may or may not be problematic depending on work conditions and the age of workers. Just as campaigns for fair and ethical work conditions that distinguish unethical (e.g. child labour) from other labour we distinguish problematic (unequal) labour from equal labour. We define an inequality footprint as the Gini index of the workforce that is directly and indirectly required to satisfy the consumption of a given population. In this way we cast income inequality into the same footprint accounting framework as scarce water, GHG emissions and unethical work practices. These accounts may then be used to study the correlative or causal relationship between consumption and inequality.

9.3 Methodology and Data Sources

9.3.1 Data Sources

This section describes the data foundation of this work. We use three databases covering 187 countries:

1. Employment: the International Labour Organization's LABORSTA database (ILO 2012), and the United Nations System of National Account UNSNA-Official Country database (UNSD 2011);
2. Income and global inter-industry transactions data: the Eora Multi-Regional Input-Output (MRIO) database (Eora 2012; Lenzen et al. 2012a, b);
3. Gini index: the Standardized World Income Inequality Database (SWIID) version 3.1 (Solt 2009), and the Organization for Economic Co-operation and Development (OCED) database (http://stats.oecd.org/) provide Gini indices referring to both before-tax and after-tax income. In this paper we have used the after-tax Gini index data. Also we including Gini index database from the World Bank (http://data.worldbank.org/).

In essence, we use data on employment, income and Gini indices to construct income distributions for the 187 countries in our study (see A.1). These income distributions are then cast into the shape of a so-called national satellite account (see A.3) that accompanies global inter-industry (MRIO) transactions data. In unison, the MRIO data and income distribution satellite enable tracing economic activity in one location to income distributions in other locations around the world. This is explained in detail in the remainder of this section.

9.3.2 Basic Input-Output Theory

This work is concerned with enumerating employment and inequality footprints for the world's economies. As in previous studies on carbon footprints (Hertwich and Peters 2009; Peters 2010), water footprints (Feng et al. 2011), and biodiversity footprints (Lenzen et al. 2012c), we apply the method of economic input-output (IO) analysis (Leontief 1986).

The centrepiece of any IO analysis is an assembly of three matrices: one $N \times N$ intermediate transactions matrix $\mathbf{T}$ with elements T_{ij} that represent monetary amounts intermediate demand from supplying economic sectors[5] $i = 1, \ldots, N$ into using sectors $j = 1, \ldots, N$; one $K \times N$ value added matrix $\mathbf{v}$ with elements v_{kj} that represent monetary amounts of primary input from value-added categories[6] $k = 1, \ldots, K$ into using sectors $j = 1, \ldots, N$; and one $N \times M$ final demand matrix $\mathbf{y}$ with elements y_{jm} that represent monetary amounts of final demand from supplying economic sectors $i = 1, \ldots, N$ into final demand categories[7] $m = 1, \ldots, M$.

This assembly is a balanced financial account in a sense that total uses $\mathbf{x} = \mathbf{T1}^N + \mathbf{y1}^M$ equal total supply $\mathbf{x}' = \mathbf{1}^k\mathbf{v} + \mathbf{1}^N\mathbf{T}$, where $\mathbf{1}^N = \{1, 1, \ldots, 1\}$ etc. are suitable summation operators, and where the prime (′) symbol denotes transposition. Setting $\mathbf{T1}^N = \mathbf{Ax}$, we find the fundamental input-output identity $\mathbf{x} = \mathbf{Ax} + \mathbf{y1}^M$, where $\mathbf{I}$ is a $N \times N$ identity matrix, $\mathbf{A}$ is the input coefficients matrix, and is the famous Leontief inverse.

9.3.3 Extended Input-Output Analysis

It was always Leontief's intention to put IO analysis to use for solving environmental and social problems (Leontief and Ford 1970; Leontief and Duchin 1986). To this end, the input-output account assembly ($\mathbf{T}$, $\mathbf{v}$, $\mathbf{y}$) is supplemented by a $1 \times N$ environmental or social satellite account $\mathbf{Q}$ with elements Q_{1j} describing amounts of some environmental or social variable (for example energy, emissions, employment) associated with (used by, emitted by) economic sectors $i = 1, 2026, N$. Setting $\mathbf{Q1}^N = \mathbf{qx}$, we find environmental or social accounting identities as $Q = \mathbf{Q1}^N = \mathbf{q}$, where Q represents the economy-wide total of the satellite account $\mathbf{Q}$. The vector $\mathbf{q}$ holds so-called intensity coefficients (for example energy intensities, employment intensities) that describe the amount of the satellite variable associated with (used, emitted per) one unit of total use. In contrast, the multipliers describe the amount of the satellite variable associated with one unit of final demand. In addition to the so-called direct effects covered by $\mathbf{q}$, the multipliers $\mathbf{m}$ include all indirect flow-on effects rippling throughout the complex supply-chain network of the entire economy, as described by the Leontief inverse. The environmental and social accounting identities are the basis of the widely used Leontief demand-pull model, which interprets Q as the total environmental or social inputs required to satisfy a final demand bundle $\mathbf{y}$. Therefore, Q is also referred to as an environmental or social footprint (depending on the nature of the satellite variable).

Whilst emissions satellite accounts are the basis of many carbon footprint studies (Hertwich and Peters 2009; Peters 2010), the satellite accounts $\mathbf{Q}$ used in this work

---

[5]For example agriculture, forestry, fishing, mining, manufacturing, utilities, trade, transport, or services.

[6]Wages and salaries, gross operating surplus, and net taxes on production.

[7]Household consumption, government final consumption, gross fixed capital expenditure, and changes in inventories.

are employment accounts **E** in units of full-time equivalent person-years, and wage and salary income accounts **W** in units of US$. These accounts were constructed at a detail of 187 countries (see A.1) with a combined 15,909 sectors in order to complement a matching Multi-Region Input-Output (MRIO) framework (**T**, **v**, **y**) of the world economy (Lenzen et al. 2012a).[8] Whilst the income satellite account **W** is based on a multitude of data sources (see Lenzen et al. 2012a), the employment satellite account **E** is based predominantly on data published by the ILO (2012). Further details on the construction procedures followed are given in A.3.

9.3.4 Inequality Satellite Accounts

The employment and income accounts **E** and **W** can be used to determine employment and income footprints of nations (Los et al. 2012). In order to construct satellite accounts on income *inequality*, data on employment and income have to be collected, and assembled into a separate satellite account, for a number of income classes. Such detailed data are hard to come by for most countries as a whole,[9] let alone for a complete suite of world nations and broken down into economic sectors as represented in the MRIO database that we use. We have therefore devised a strategy to estimate distributions of income across C income classes for individual industry sectors in individual countries from three data items: (a) the country's Gini index γ, (b) the sector's total income payments for employees I_{tot}, and (c) the sector's total workforce P_{tot}.

Income inequality is typically depicted in Lorenz curves (Fig. 9.1). Cumulative income $I(c)$ at income class c (normalised to 100%)

$$I(c) = \frac{1}{I_{\text{tot}}} \int_0^c i(c)dc \tag{9.1}$$

is plotted against cumulative workforce (equally normalised).

$$P(c) = \frac{1}{P_{\text{tot}}} \int_0^c p(c)dc. \tag{9.2}$$

In the case of perfect income equality, cumulative income increases proportionally with cumulative workforce, and the Lorenz curve is a diagonal connecting the origin with the point {1, 1} (dashed line). Any degree of income inequality will

[8]For further information on MRIO frameworks and MRIO analysis please consult Leontief and Strout (1963), Miller and Blair (2010), and Murray and Lenzen (2013).

[9]For example the Household Expenditure Surveys of Brazil (IBGE 2010), and Australia (ABS 2011).

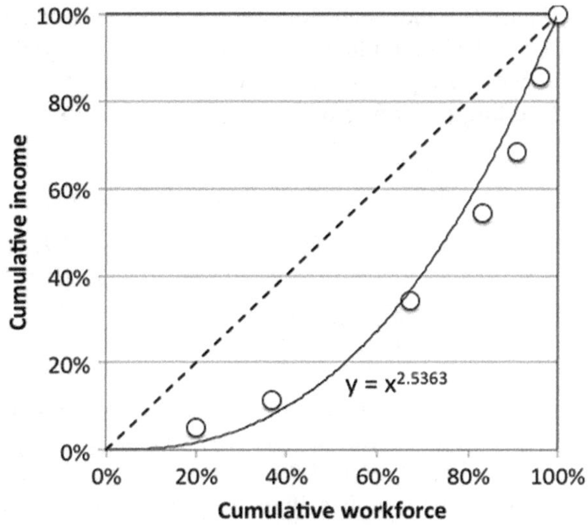

Fig. 9.1 Lorenz curve for Brazil's income distribution in 2009 *Notes Circles* data from IBGE (2010); *solid line* and regression equation: power function approximation from Eq. 9.3; *dashed diagonal line* Lorenz curve for complete income equality. Further examples about the quality of the power function fit can be found in A.4

see cumulative workforce increase more rapidly that cumulative income, thus leading to a convex Lorenz curve (circle markers in Fig. 9.1).

$$I(c) = f[P(c)]. \tag{9.3}$$

This Lorenz curve can be approximated by a power function (solid line in Fig. 9.1).

$$f = P(c)^{\xi}. \tag{9.4}$$

Total income equality corresponds to $\xi = 1$, and $\xi \to \infty$ leads to increasing inequality. A power function also ensures that all Lorenz curve representations pass through the origin $\{0, 0\}$ and the point $\{1, 1\}$.

The Gini inequality index γ can be calculated as the ratio of the area between the diagonal and the Lorenz curve, and the area below the diagonal:

$$\gamma = \frac{1/2 - \int_0^C I(c)}{1/2} = 1 - 2\int_0^C I(c) = 1 - 2\int_0^C P(c)^{\xi} = 1 - \frac{2}{\xi + 1} P(c)^{\xi+1}\Big|_0^C. \tag{9.5}$$

Since $P(0) = 0$ and $P(C) = 1$ per definition in Eq. 9.2, we can evaluate the integral, and find the relationship between ξ and γ as

$$\gamma = 1 - \frac{2}{\xi + 1} \Leftrightarrow \xi = \frac{1 + \gamma}{1 - \gamma}. \tag{9.6}$$

Figure 9.1 shows an example for an approximation of a Brazilian Lorenz curve with $\xi = 2.51$ (solid line), based on data (circles) yielding $\gamma = 0.43$. We have added evidence in support of the power function approximation in A.4.

Inserting Eq. 9.6 into Eq. 9.3, we find the income distribution

$$\begin{aligned}\frac{p(c)}{P_{\text{tot}}} &= \frac{\mathrm{d}}{\mathrm{d}c}[P(c)] = \frac{\mathrm{d}}{\mathrm{d}c}\left[I(c)^{1/\xi}\right] = \frac{\mathrm{d}}{\mathrm{d}c}\left[\left(\frac{1}{I_{\text{tot}}}\int_0^c i(c)dc\right)^{\frac{1-\gamma}{1+\gamma}}\right] \\ &= I_{\text{tot}}^{-\frac{1-\gamma}{1+\gamma}}\frac{1-\gamma}{1+\gamma}\left(\int_0^c i(c)dc\right)^{\frac{1-\gamma}{1+\gamma}-1} i(c) = I_{\text{tot}}^{-\frac{1-\gamma}{1+\gamma}}\frac{1-\gamma}{1+\gamma}\left(\int_0^c i(c)dc\right)^{-\frac{2\gamma}{1+\gamma}} i(c).\end{aligned} \tag{9.7}$$

Equation 9.7 can be evaluated by choosing income intervals $i(c)$, and calculating corresponding workforce fractions. Choosing a large number C of intervals will increase the resolution of the approximation. Figure 9.2 shows an example for an approximation of a Brazilian Lorenz income distribution curve, based on data (circles) yielding $\gamma = 0.43$.

The average wage $w(c)$ for each income class c can be determined via

$$w(c) = i(c)/p(c) \tag{9.8}$$

Assuming that an income class satellite would identify $L \ll C$ classes according to wage intervals $w_l < w(c) \leq w_{l+1}$, and assuming that one satellite class $l \in \{1,$

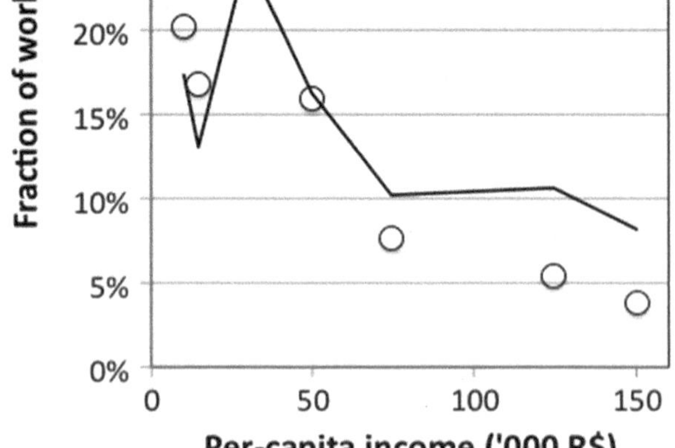

Fig. 9.2 Income distribution for Brazil in 2009. *Notes Circles* data from IBGE (2010); *solid line* power function approximation from Eq. 9.7

..., $L\}$ contains n_l such wage classes $\{c1, ..., nl\}$, the income class satellite account **Q** simply reads

$$Q(l) = \sum_{\substack{c\,with \\ w_l < w(c) \leq w_{l+1}}} i(c) = \sum_{n=1}^{n_l} i(c_n). \tag{9.9}$$

The workforce in each satellite class l is then taken from Eq. 9.7 as

$$R(l) = \sum_{\substack{c\,with \\ w_l < w(c) \leq w_{l+1}}} p(c) = \sum_{n=1}^{n_l} p(c_n). \tag{9.10}$$

The average wage in each satellite class l is

$$\bar{w}(l) = Q(l)/R(l), \tag{9.11}$$

with $w_l < \bar{w}(l) \leq w_{l+1}$. Repeated for all sectors in the MRIO database, this procedure will yield two satellite accounts **Q** and **R** sized $C \times N$, one for employment by class, and one for income by class.

9.3.5 Inequality Footprints

Inequality footprints of nations can now be calculated subjecting the income and employment class accounts to the Leontief-type demand-pull terms of income inequality would then proceed as follows. Assume a $N \times 1$ vector $\boldsymbol{\psi}_\chi$ describing a final demand bundle originating from country χ, the inequality footprint of which is to be evaluated. The impact of $\boldsymbol{\psi}_\chi$ in terms of income earned in economic sectors j in various classes c is $\boldsymbol{\theta} = \{\theta_{cj}\} = \mathbf{q}(\mathbf{I} - \mathbf{A})^{-1}\boldsymbol{\psi}_\chi$, where $\mathbf{q} = \mathbf{Q}\hat{\mathbf{x}}^{-1}$ holds income by sector and by class per unit of total use, and where $\hat{\mathbf{x}}$ is a diagonal matrix holding total use. The impact of $\boldsymbol{\psi}$ in terms of employment in various classes is $\boldsymbol{\rho} = \mathbf{r}(\mathbf{I} - \mathbf{A})^{-1}\boldsymbol{\psi}_\chi$, where $\mathbf{r} = \mathbf{R}\hat{\mathbf{x}}^{-1}$ holds employment by sector and by class per unit of total use. In other words, $\boldsymbol{\theta}$ and $\boldsymbol{\rho}$ are the income and the workforce, respectively, broken down by economic sector and by class, that are required directly and indirectly to satisfy the final demand bundle $\boldsymbol{\psi}_\chi$, or in other words the *income and employment footprints* of the final demand bundle $\boldsymbol{\psi}_\chi$. The average wage $\bar{w}$ paid to this workforce in sector j is $\bar{w} = \sum_c \theta_{cj} / \sum_c \rho_{cj}$.

The Gini index γ characterising this global workforce is

$$\gamma_\chi = \frac{\frac{\sum_{cj}\theta_{cj}\sum_{cj}\rho_{cj}}{2} - \sum_{cj}\theta_{cj}\rho_{cj}}{\frac{\sum_{cj}\theta_{cj}\sum_{cj}\rho_{cj}}{2}} \tag{9.12}$$

γ_χ includes people working in country χ and in other countries.[10] It is possible to isolate the inequality effect of the domestic final demand bundle ψ_χ on particular countries δ by only considering the income and workforce needed from the sectors $j \in \delta$ belonging to those countries δ:

$$\gamma_{\delta\chi} = \frac{\frac{\sum_{c,j\in\delta}\theta_{cj}\sum_{c,j\in\delta}\rho_{cj}}{2} - \sum_{c,j\in\delta}\theta_{cj}\rho_{cj}}{\frac{\sum_{c,j\in\delta}\theta_{cj}\sum_{c,j\in\delta}\rho_{cj}}{2}} \tag{9.13}$$

The *inequality footprint* of a nation can be defined as a weighted sum over countries of production origin

$$\gamma_{\mu\chi} = \frac{\frac{\sum_{c,j\notin\chi}\theta_{cj}\sum_{c,j\notin\chi}\rho_{cj}}{2} - \sum_{c,j\notin\chi}\theta_{cj}\rho_{cj}}{\frac{\sum_{c,j\notin\chi}\theta_{cj}\sum_{c,j\notin\chi}\rho_{cj}}{2}} \tag{9.14}$$

The *inequality footprint of imports* can be defined as a weighted sum over countries of imports origin

$$\bar{\gamma}_\chi = \sum\nolimits_\delta w_{\delta\chi}\gamma_{\delta\chi} / \sum\nolimits_\delta w_{\delta\chi} \tag{9.15}$$

The origin country weights $w_{\delta\chi}$ could be set to the income footprints $\theta_{\delta\chi}$ or employment footprints $\rho_{\delta\chi}$, or to the product $\theta_{\delta\chi}\rho_{\delta\chi}$ of both.[11] In general the more unequal the selection of import origins, the higher the inequality footprint of imports.

[10]Refers to the inequality within the entire world, and therefore relate to the between-country perspective of inequality described by the 'World Inequality' concept in Milanovic (2005). In general the more unequal the combination of import origins, the higher the global inequality footprint. This perspective is equivalent to. Milanovic (2005) also offers a method for decomposing global inequality into within-country, between-country, and overlapping effects.

[11]Compare with Part A in Eq. 9.1 in Milanovic (2005).

9.3.6 Qualifications

To our knowledge this is the first time that an income inequality indicator (through the Gini index measure) has been combined with an input-output calculus. Nevertheless, the implementation of the method we have so far described has a number of shortcomings.

First, as we described above, income distributions are largely unavailable for most countries, and therefore we needed to construct income distributions from Gini indices by fitting power functions. These power function fits will necessarily deviate from "true" income distributions, but as we have shown in the example for Brazil, and in more examples listed in A.4, such deviations are likely to be small.

Second, Gini indices are not available as continuous time series for all countries. In the absence of continuous information, we interpolated the Gini indices for missing years on the basis of those for neighbouring years. We treat missing values at the beginning and the end of our period of analysis by setting them equal to the data available for the closest year. Large gaps exist only for countries with a combined small fraction of GDP, and associated errors are probably small. For Gini index availability see A.2.

Third, data for Gini indices, income distributions, or Lorenz curves are almost impossible to locate for individual industry sectors, therefore we needed to assume that the national Gini index applies to all sectors of the economy. This procedure will lead to allocation errors in cases where inequality varies significantly between sectors, *and* where a country predominantly exports those products made by sectors with significantly above- or below-average inequality (for example, soybean in Brazil as opposed to other sectors such as petroleum oils).

9.4 Results

We find that the inequality footprint of countries differs substantially from their within-country inequality (Gini index) (Fig. 9.3). Countries occupy top ranks in Fig. 9.3 if their inequality footprint is significantly larger than their within-country inequality. This can be either because their imports come from unequal countries (for example Japan which imports from China, Thailand, Russia, etc. see Table 9.1), or their own country's society is very equal (for example Sweden which occupies a rank similar to that of Japan but for different reasons see Table A.6 for supporting data). The opposite pattern holds for the bottom-ranked countries. Their inequality footprint is significantly smaller than their within-country Gini index. That means either their imports come from equal countries (for example in the case of Russia), or their own country's society is very unequal (for example in the case of South Africa) (see Table A.6). In addition to top- and bottom-ranked countries we include France, Korea, United Kingdom, United States, Thailand, and China

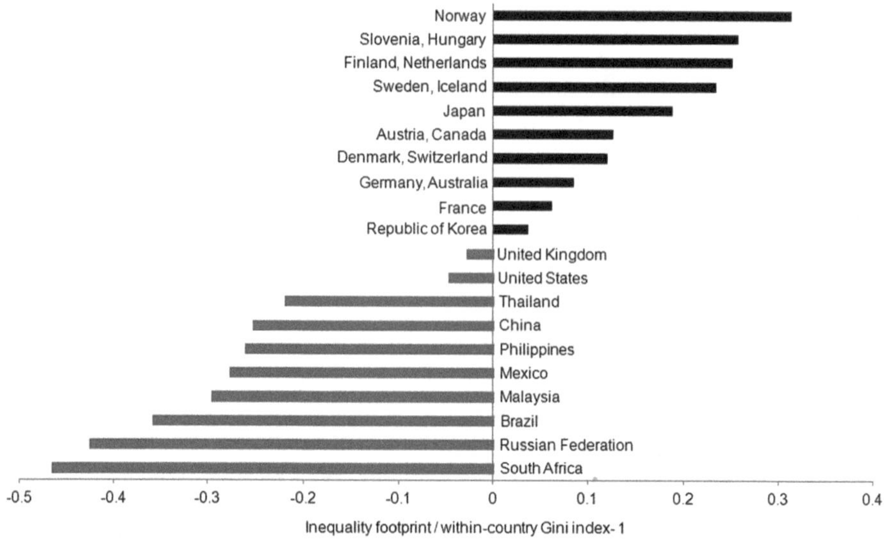

Fig. 9.3 The world's *top* and *bottom* countries in terms of the disparity between their inequality footprint and within-country inequality. *Notes* the horizontal axis depicts the departure from 1 of the ratio between inequality footprint and within-country Gini index

because of their significant share of the world's Gross Domestic Product (GDP) (for more information about rankings see Table A.5).

Scandinavian countries have the largest disparity between their inequality footprint and their within-country inequality (Fig. 9.3 and Table A.6). Further investigation of the commodities that are being traded reveals that inequality is often hidden from the final consumer because of complex supply chains that stretch across multiple countries and producers. The supply chains ending in electronic goods often originate in export-oriented industries in developing countries producing inputs such as electronic components, chemicals, fertilizers, minerals, and agricultural commodities characterised by high levels of inequality (Table 9.1). For example, the consumption of a mobile phone in Norway may require labour from China and Thailand to assemble phone components, which in turn requires electronic circuit manufacturing in the Philippines and Malaysia, which in turn relies on Russian petroleum oils. Despite the origins of Scandinavian imports often being unequal societies that does not mean the consumption in Norway, for example *causes* inequality in another country. It could even be that the consumption decreases inequality. We may simply say that through their consumption Norwegians are *implicated* in or associated with the situation of unequal income distributions (compare Sect. 9.2.3). Although Norwegian society itself maintains high levels of equality through active pursuit of re-distributive tax, unemployment and social benefits policies, employment protection legislation for workers, and reduced taxes on labour for low-income workers (OECD 2011b).

Table 9.1 This table ranked list of countries as in Fig. 9.3 but with detail on inequality-implicated commodities and the labour embodied in imports from countries that have a Gini index above 0.4

Country	% above 0.40	Inequality-implicated commodities (country embodied labour ('000 FTE) implicated-commodity)	% 0.35–0.4	% 0.3–0.35	% Less 0.3
Norway	34	CHN 128 mob; RUS 82 po; THA 75 tel; PHL 40 ec; BRA 38 sb; MDG 20 va	16	27	23
Slovenia	33	RUS 43 ng; CHN 26 mob; THA 13 ac	13	31	23
Hungary	38	RUS 157 po, gas; CHN 70 mob; THA 27 ec	10	23	28
Finland	43	RUS 195 po; CHN 86 pt, mob; THA 45 tra	13	20	24
Netherlands	36	CHN 2150 mob; RUS 252 po; THA 202 mob; BRA 182 sb; PHL 175 ec; MDG 78 ff; UGA 70 tr	20	25	19
Sweden	29	RUS 145 po; CHN 135 mob; THA 80 mob; PHL 45 ec	16	24	31
Iceland	39	RUS 11 po; THA 9 dum; CHN 8.8 com	18	22	20
Japan	51	CHN 21,500 mob; THA 2360 med; RUS 2134 po; PHL 2057 ec; MDG 1161 va; MYS 728 ch; BRA 500 cop; LKA 418 sf	23	19	6
Switzerland	55	RUS 3000 po, jew; PHL 668 ec; CHN 407 com; THA 379 cw; LKA 323 jew; ECU 266 coc, ros; UGA 245 tob; MDG 145 coc	15	15	15
Austria	30	CHN 155 mob; RUS 139 po; THA 73 wd; MDG 69 ff, va; PHL 50 ec; BRA 39 ff	15	23	32
Denmark	32	CHN 125 com; RUS 101 po; THA 70 mt	16	26	27
Canada	39	CHN 3251 mob; PHL 295 tx; MEX 275 gen; RUS 174 po; THA 180 mob; MDG 144 va; BRA 147 med, tra	35	16	9
Germany	43	CHN 8490 mob, clo; RUS 3184 po; MDG 683 va; BRA 605 pu; PHL 575 ec, sc; THA 591 jew, mob	15	25	17
Australia	42	CHN 4884 com, med; THA 488 med, mob; MYS 313 scr; PHL 364 ec; PNG 120 po; RUS 122 po; BRA 108 io	23	27	9
France	38	CHN 4894 com, pm; MDG 1173 ff, va; RUS 776 po; MAR 608 clo, veg; BRA 366 pu; THA 360 ac	20	27	15
Republic of Korea	41	CHN 974 mob, pm, clo; PHL 381 ec, ff; THA 377 tel; RUS 298 po, coal; MYS 155 ec	25	20	14
United Kingdom	38	CHN 7284 mob, tel; RUS 896 po; THA 879 ec; PHL 820 ec; MYS 237 tel	20	23	20
United States	55	CHN 27,178 tra, rubb, med, clo; MEX 4853 mot, cof; PHL 2010 ec; RUS 1428 po, jew; BRA 1345 tra, tob; THA 1,368 tel, jew; MDG 896 va	10	17	18

(continued)

Table 9.1 (continued)

Country	% above 0.40	Inequality-implicated commodities (country embodied labour ('000 FTE) implicated-commodity)	% 0.35–0.4	% 0.3–0.35	% Less 0.3
Thailand	27	CHN 261 com, mob; PHL 130 ec, tra; RUS 95 po, pu; MYS 105 ec	40	19	13
China	32	THA 1129 rubb, mob; RUS 1040 wd, io; MDG 720 wd; PHL 741 ec; MYS 463 wd, rubb, ec; BRA 313 io, po	25	26	18
Philippines	30	THA 149 rc; CHN 69 pg; RUS 34 io, po; MYS 23 ec, ec	31	26	13
Mexico	41	CHN 281 com; RUS 172 po; BRA 157 tra; THA 98 pt; GTM 83 sug; PHL 78 ec; CHL 77 ff; MDG 75 nb; MYS 41 ec	34	16	10
Malaysia	38	THA 638 rubb, mob; CHN 289 ec; PHL 227 ec; RUS 101 io, po; MDG 49 clov	29	24	9
Brazil	52	RUS 585 po; CHN 413 tel; ARG 362 mot; BOL 349 pg; THA 178 med, tra; PRY 163 ma; PHL 166 ec; MEX 111 ref; VEN 101 po, coke	18	17	13
Russia Federation	12	UKR 2293 rw, io; BLR 1180 n.e.s.; TJK 595 cot, ff; KAZ 530 io; AZE 309 ff; LTU 52 dp, ff	18	20	50
South Africa	44	CHN 129 com, mob; ZMB 126 cot, cop; THA 72 rc; RUS 55 cop; BRA 38 mv; MDG 36 clov	20	20	16

Notes The inequality footprint is broken down into contributions from trade partners with a Gini index of above 0.4, 0.35–0.4, 0.3–0.35, and less 0.3. *ARG* Argentina, *AZE* Azerbaijan, *BLR* Belarus, *BOL* Bolivia, *BRA* Brazil, *CHL* Chile, *CHN* China, *ECU* Ecuador, *KAZ* Kazakhstan, *LKA* Sri Lanka, *LTU* Lithuania, *MDG* Madagascar, *MEX* Mexico, *MYS* Malaysia, *PHL* Philippines, *PRY* Paraguay, *RUS* Russia, *THA* Thailand, *TJK* Tajikistan, *UGA* Uganda, *UKR* Ukraine, *VEN* Venezuela, *ZMB* Zambia, *ac* air conditioner, *ch* wood charcoal, *clo* clothes, *clov* cloves, *coc* cocoa, *cof* coffee, *cop* copper, *cot* cotton, *cw* clocks and watches parts, *dp* dairy products, *dum* dumpers, *ec* electronic circuits, *ff* fresh fruits and juices, *gen* electric generators, *io* iron ores, *jew* jewellery, *ma* maize, *med* medical articles and instruments, *mob* mobile, *mot* electric motors and it's parts, *mt* canning meat, *n.e.s.* not elsewhere specified, *nb* niobium ore, *ng* natural gas, *pg* petroleum gas, *pm* printing machine, *po* petroleum oil, *pt* part of telephone, *pu* chemical wood pulp, *rc* milling rice, *ref* refrigerators, *ros* roses, *rubb* natural rubber, *rw* railway parts, *sb* soya bean, *sc* solar cell, *scr* monitors and projectors, *sf* seafood, *sug* cane or beet sugar, *tel* telephone, *tob* tobacco, *tr* live tree, *tra* tractors parts and accessories, *tx* textiles, *va* vanilla, *veg* vegetables, *wd* wood in rough

Many developed countries have an inequality footprint that is higher than their within-country inequality largely because of their imports from more unequal developing or transition economies. Notable exceptions are the United States and the United Kingdom (Fig. 9.3). These exceptions are due to their own country's society being unequal (see Table A.6). Whilst their inequality footprint is not much different from that of Switzerland or the Netherlands, their own Gini indices of 0.33 and 0.36 are considerably above those of most European countries (Gini index <0.30). Seventy million full-time equivalent workers from outside the USA support the lifestyle of US citizens (this is equivalent to 50% of the total workforce in the

USA for 2010), approximately 40 million of them are from countries with high inequality (see examples in Table 9.1). The main imports of developed countries (for example US) are electronics and clothes, raw materials and natural resources, and manufactured parts that are used as an intermediate input into making other products (Table 9.1). Examples for the latter category are: motor vehicle parts from Mexico (where about 5 million laborers are working for US) exported for input into car production in the US; electronics from China (about 27 million working for US); electronic integrated circuits from the Philippines (about 2 million working for US) which are also used as an input into electronic production in the US. Despite the high amount of labour embodied in imports from unequal countries and a large inequality footprint, the within-country Gini index of the United States is higher than its total inequality footprint. Some economists attribute (Carnevale and Rose 2011) the rising income inequality in the US to the large gap in wages between college-educated and high school workers. Another factor affecting increasing inequality in the US is the large disparity in income distribution between skilled and unskilled workers and the low level of income redistribution. Further, the opening up of trade in the United States could affect inequality, for example through the North American Free Trade Agreement (NAFTA) between the United States, Canada, and Mexico in 1994. This treaty increased income inequality in the US because the unemployment rate increased when unskilled jobs moved from.

The United States (high-wage) to Mexico (low-wage). The rise in within-country inequality in the United States is coupled with a rise in the inequality footprint in the past two decades (see Fig. A.7). Figure 1 in A.7 shows a small decrease in inequality after 2003 with an increase following the financial crisis and a slight decrease again in the past four years during Barack Obama's presidency. This is perhaps because his election campaign included amending the NAFTA and its outsourcing of jobs to other countries.

Also in the United Kingdom within-country inequality is higher than its total inequality footprint. Our results show that the inequality footprint increased after the Tony Blair era with a decrease in 2003 and then increased again after the global financial crisis and recession in the United Kingdom after 2008 (see Fig. A.7). The increasing inequality footprint after the recession may be due to rising commodity prices in the United Kingdom after 2008 (Tang 2008). Other researchers (UNCTAD 2012) attribute this rising inequality to a decline in the progressivity of the personal income tax schedule of the overall tax system in the United Kingdom and the United States and others (Irvin 2011) attribute the increase in within-country inequality in the United Kingdom and United States to the Reagan–Thatcher era which gave the green light to financial deregulation and dependence on foreign labour (labour embodied) through trading activities.

Russia occupies an unusual position at the bottom of the rank, because both its inequality footprint and its own Gini index are lower than those of similarly ranked countries such as Brazil and South Africa (see Table A.6). This is because approximately 62% of Russia's total share of embodied labour arrives from the Commonwealth of Independent States where inequality is low (Table 9.1). Most of its imported commodities are from agriculture or are raw materials (Table 9.1).

Figure 1 in A.7 shows that the inequality footprint and within-country inequality increased in Russia after the collapse of the Soviet Union. The increase in the inequality footprint occurred because either within-Ukraine inequality increased in the 1990s (our findings showed that the highest labour embodied in the past two decades was from Ukraine (about 24% in 2010) [SWIID version 3.1 (Solt 2009)] or Russia's other trading partners were from unequal countries. Also, the figure shows a decrease in the inequality footprint in 2000s, which is either because of the decreasing within-Ukraine inequality [SWIID version 3.1 (Solt 2009)] in the 2000s or the economic crisis at the end of the 1990s in Russia.

Russia's low inequality footprint is reflected in its recent history. After the collapse of the former Soviet Union, tight economic links amongst the newly independent republics and the former Eastern Bloc continued (see data in The Observatory of Economic Complexity 2012). Most of Russia's current trade partners achieved or maintained low inequality, either by keeping parts of the Soviet-style social systems such as pension benefits, and by avoiding economic reforms such as privatization (International Monetary Fund, 2006), or, like Lithuania, Eastern Germany and Poland, by joining the European Union. On the other hand, Russia experienced a significant increase in within-country inequality due to the privatization of state-owned enterprises and inequality in wages within the private sector between men, women, and older workers (Brainerd, 1998). However, because of the equality of its trading partners its inequality footprint remains low.

Investigating the main imported commodities between developing countries (Table 9.1) shows that, for example, China and Thailand import chiefly raw materials and intermediate goods. China imports raw materials such as natural rubber from Thailand and Malaysia, iron ore from Russia and Brazil, and wood in rough from Madagascar and intermediate products (electronic integrated circuits from the Philippines and Malaysia). Thailand also imports raw materials (or natural resources) from Russia (petroleum oil) and intermediate goods from Philippines and Malaysia (electronic integrated circuits). By recognizing the flows of embodied labour (or employment footprint) and the commodities traded between developing (or maybe developed) countries we provide information that be used can support a policy agenda to address inequality within those countries (more details in conclusion).

The remaining countries with an inequality footprint higher than their own Gini include Brazil and South Africa, which are both characterised by high levels of inequality, probably because both lack policies to provide a comprehensive social security net, and both have experienced a rise in the wages gap between skilled and unskilled labour in rural and urban regions (World Bank 2000; Kingstone 2011). In Brazil, the main agricultural products that may be implicated in within-country inequality are soya bean and fruit juices. Labour embodied in these commodities is exported to the Netherlands, Norway, and Austria, thus increasing the inequality footprints of those countries. Ores and wood pulp from Brazil also play a large part in increasing inequality footprints of importing countries (Table 9.1). In South Africa, inequality has historical roots going back to the beginning of the last century and apartheid when 10% of the population (i.e. white) held 50% of the total share of income (World Bank 2000) while the other racial groups (coloured 8%, Indian/Asian 2%, and African 80%) held 10%, 32%, and 8% respectively. During the last two decades inequality has declined between racial groups through the inclusion of blacks in the mining industry, which has allowed African workers to be classified as employees (Jenkins and Thomas 2011). However inequality has risen within racial groups especially within African groups because of the large number of uneducated, unskilled workers and the small number of more highly paid educated workers (Jenkins and Thomas 2011). Despite its geographical location surrounded by highly unequal countries, such as Namibia, Botswana and Madagascar, the inequality footprint of South Africa is equal to the footprint of some developed countries (Table 9.1). This is probably because embodied labour imported from countries that have a within-country Gini index above 0.4 is less than from countries that have a within-country Gini index below 0.4 (Table 9.1).

Our results show that the highest inequality footprint in the world is that of Namibia (0.5), because approximately 81% of total labour embodied in imports was from countries that have high levels of inequality (for example South Africa 65%, Russia 6%, and China 3%). However Fig. 9.3 does not show Namibia in the bottom-ranking countries because its own inequality is also the highest in the world (about 0.64 according to the World Bank). Therefore there is little difference between its within-country inequality and its inequality footprint.

The inequality footprint can add to our understanding of within-country inequality tracking implicated-commodities moving around the world. It also facilitates the distinguishing of problematic unequal labour from equal labour worldwide just as we already distinguish 'scarce water' from 'non-scarce water' (Hoekstra and Chapagain 2007) and ethical labour from unethical labour such as child labour (Basu and Zarghamee 2009). Our work provides another tool that can assist businesses, government and non-government organizations to identify areas of responsibility and take action to make the world a better place.

9.5 Conclusion

Both developing countries (producers) and developed countries (consumers) may suffer from inequality because of trade activities. Our findings showed that the major commodities associated with wage inequality during production were agricultural and electronic commodities. Most developing countries who deal in these commodities have an inequality footprint less than their within-country Gini index. Among developed countries only the United States and the United Kingdom have within-country inequality higher than their inequality footprint. The United States shows an increasing inequality footprint during the Clinton era and especially after the NAFTA treaty came into force and then decreasing inequality footprints after 2003 and increasing again after the financial crisis of 2007 and slightly decreasing during Barack Obama's presidency. The United Kingdom shows that the inequality footprint increased after the Tony Blair era with a decrease in 2003 and then increasing again after the global financial crisis and recession in the United Kingdom after 2008. Both the United States and the United Kingdom showed a decrease in their inequality footprint after 2003 and this may be related to the war in Iraq and their relationships to Russia and China. Russia has demonstrated the lowest inequality footprint in the Commonwealth of Independent States and in the world.

Inequality is being addressed by national governments, global organisations, non-government organisations and the business world. The UN's Human Development Report 2010 concludes that equity and poverty must be at the forefront of policy development and conflict resolution, this includes global equitable trade and investment rules and policies to address climate change. Non-government organisations like *Fair Trade* are committed to alleviating poverty around the world in an effort to make life more equitable.[12] The Inclusive Business Dialogue held at the 2010 UN Summit on the Millennium Development Goals was attended by over 200 delegates from the private sector, government and donors who discussed the practice of 'inclusive business', defined as business initiatives that work for the poor at the base of the pyramid as well as for businesses. Its focus is on 'core business'—creating jobs and incomes and building skills, which are the very areas identified by the OECD as the most important factors in the building of a more equitable society (OECD 2011a). Our work supports this policy agenda by providing more detailed reliable supply chain information about the inequality footprint, labour embodied through trading activities with the rest of the world, and by providing an historical time series back to 1990. In addition we have the ability to continuously update the entire database with about one year's delay.

We began this paper with a reference to the first 1990–2015 Millennium Development Goal; we end with a comment on the post-2015 agenda. The UN's Task Team on the post-2015 UN development agenda convened to address the issue of inequalities concludes that inequalities must be addressed along with human rights, peace, security and sustainability as a cornerstone of the society that we want to live in and pass on to the next generation (UN System Task Team 2012).

Any new policy must be designed from a global perspective to reduce inequality. The UN Systems Task Team on the Post-2015 UN Development Agenda (2012) recommends that there be a specific post-2015 goal on inequality with accompanying targets and equity-weighting indicators including income distribution. This they acknowledge will require investment in data collection, analysis and use, requiring a flexible standard measurement to capture, track and reflect inequalities and provide for transparency and accountability in progress. The Task Team further linked inequality to sustainability as a cornerstone of this agenda. It is difficult for individual countries to take action on what is a global problem. One way to address this, suggested by Basu (2005), would be to create a global body to coordinate

[12]http://www.fairtrade.net/standards.html (Accessed 22/10/12).

inter-country anti-poverty and anti-inequality policies, operating in the same way as the WTO helps mitigate problems in trade and the UNEP in environmental problems (Basu 2005). This is where the inequality footprint can play a role. It provides a map of the movement of embodied labour across countries and detailed information about which commodities are the most labour-intensive and may be implicated in within-country inequality.

The inequality footprint can add to our understanding of within-country inequality tracking implicated-commodities moving around the world. It also facilitates the distinguishing of problematic unequal labour from equal labour worldwide just as we already distinguish 'scarce water' from 'non-scarce water' (Hoekstra and Chapagain 2007) and ethical labour from unethical labour such as child labour (Basu and Zarghamee 2009). Our work provides another tool that can assist businesses, government and non-government organizations to identify areas of responsibility and take action to make the world a better place.

References

ABS. (2011). *Household expenditure survey—Detailed expenditure items.* ABS Catalogue No. 6535.0. Canberra, Australia: Australian Bureau of Statistics.

Atkinson, A. B., Piketty, Thomas, & Saez, Emmanuel. (2011). Top incomes in the long run of history. *Journal of Economic Literature, 49*(1), 3–71.

Basu, K., (2005). Globalisation, poverty and inequality. Helsinki, UNU-WIDER. Research Paper No. 2005/32.

Basu, K., & Zarghamee, H. (2009). Is product Boycott a good idea for controlling child labor? A theoretical investigation. *Journal of Development Economics, 88*(2).

Borjas, G. J., & Ramey, V. A. (1995). Foreign competition, market power and wage inequality. *The Quarterly Journal of Economics, 110*(4), 1075–1110.

Brainerd, E. (1998). Winners and losers in Russia's Economic Transition. *The American Economic Review, 88*(5).

Burke, J. (2000). *Child labour scandal hits Adidas.* Retrieved January 3, 2013 from theguardian| TheObserver: http://www.guardian.co.uk/uk/2000/nov/19/jasonburke.theobserver

Carnevale, A., & Rose, S. J. (2011). *The undereducated American.* Retrieved June 26, 2011 from http://cew.georgetown.edu/undereducated

Chapple, S., Forster, M., & Martin, J. P. (2009). *Inequality and well-being in OECD countries: What do we know?* Busan, Korea: OECD.

Collier, P. (2007). *The bottom billion: Why the poorest countries are failing and what can be done about it.* Oxford, UK: Oxford University Press.

Denisova, I. (2012). *Income distribution and poverty in Russia.* OECD Social, Employment and Migration Working Papers, No. 132.

Dreher, A., & Gaston, N. (2008). Has globalisation increased inequality? *Review of International Economics, 16*(3), 516–536.

Eora. (2012). *The Eora global multi-region input-output tables.* Internet site http://www.worldmrio.com. Sydney, Australia: ISA, The University of Sydney.

Feng, K., Chapagain, A. K., Suh, S., Pfister, S., & Hubacek, K. (2011). Comparison of bottom-up and top-down approaches to calculating the water footprints of nations. *Economic Systems Research, 23,* 371–385.

Foley, S. (2009). *Walmart embroiled in child labour scandal*. Retrieved January 3, 2013 from Independent.ie: http://www.independent.ie/world-news/walmart-embroiled-in-child-labour-scandal-1930833.html

Galor, O., & Moav, O. (2000). Ability-biased technological transition, wage inequality, and economic growth. *The Quarterly Journal of Economics, 115*(2), 469–497.

Goldberg, P. K., Khandelwal, A. K., Pavcnik, N., & Topalova, P. (2010). Imported intermediate inputs and domestic product growth: Evidence from India. *The Quarterly Journal of Economics, 125*(4), 1727–1767.

Goldberg, P., & Pavcnik, N. (2007). Distributional effects of globalization in developing countries. *Journal of Economic Literature, 45*(1), 207–232.

Hertwich, E. G., & Peters, G. P. (2009). Carbon footprint of nations: A global, trade-linked analysis. *Environmental Science and Technology, 43*, 6414–6420.

HM Government. (2005). *Securing the future*. UK Sustainable Development.

Hoekstra, A. Y., & Chapagain, A. (2007). Water footprints of nations: Water use by people as a function of their consumption pattern. *Water Resource Management, 21*, 35–48.

IBGE. (2010). *Pesquisa de Orçamento Familiares*. Rio de Janeiro, Brazil, Instituto Brasileiro de Geografia e Estatística, Ministério da Fazenda.

ILO. (2012). *LABORSTA—Main statistics (annual): Employment general level, by economic activity, by occupation, by status in employment*. Internet site http://laborsta.ilo.org. Genève, Switzerland: International Labour Organization.

Inclusive Business Dialogue. (2010). *Delivering results: Moving towards scale*. Inclusive Business Dialogue. New York: UN Summit on the Millennium Development Goals.

International Monetary Fund. (2006). Republic of Belarus, Selected Issues. IMF Country Report No. 06/315.

International Monetary Fund. (2007). *World economic outlook: Globalization and inequality*. Washington, DC, USA: International Monetary Fund, Publication Services.

Irvin, G. (2011). Inequality and recession in Britain and the USA. *Development and Change, 42*(1), 154–182.

Jackson, T. (2012). *Prosperity without growth? Economics for a finite planet*. London, GB: Taylor and Francis Ltd.

Jenkins, C., & Thomas, L. (2011). The changing nature of inequality in South Africa. The United Nations University-WIDER Working Papers No. 203.

Kingstone, P. (2011). *The political economy of Latin America*. New York, NY, USA: Taylor & Francis.

Krugman, P., & Venables, A. J. (1995). Globalization and the inequality of nations. *The Quarterly Journal of Economics, 110*(4), 857–880.

Lenzen, M., Kanemoto, K., Moran, D., & Geschke, A. (2012a). Mapping the structure of the world economy. *Environmental Science and Technology, 46*, 8374–8381. doi:10.1021/es300171x.

Lenzen, M., Kanemoto, K., Moran, D., & Geschke, A. (2012b). Building Eora: A global multi-region input-output database at high country and sector resolution. *Economic Systems Research* (in press).

Lenzen, M., Moran, D., Kanemoto, K., Foran, B., Lobefaro, L. & Geschke, A. (2012c). International trade drives biodiversity threats in developing nations. *Nature, 486*, 109–112.

Leontief, W. (1986). *Input-output economics*. New York, NY, USA: Oxford University Press.

Leontief, W., & Duchin, F. (1986). *The future impact of automation on workers*. Oxford, UK: Oxford University Press.

Leontief, W., & Ford, D. (1970). Environmental repercussions and the economic structure: An input-output approach. *Review of Economics and Statistics, 52*, 262–271.

Leontief, W. W., & Strout, A. A. (1963). Multiregional input-output analysis. In T. Barna (Ed.), *Structural interdependence and economic development* (pp. 119–149). London, UK: Macmillan.

Los, B., Timmer, M. P., & de Vries, B. (2012). China and the world economy: A global value chain perspective. In *Global value chain in the 21st century: Policy implications on trade,*

investment, statistics, and developing countries. Beijing, China: Ministry of Commerce of China, WTO, UNCTAD and OECD.
Milanovic, B. (2005). *Worlds apart: Measuring international and global inequality*. Princeton, USA: Princeton University Press.
Miller, R. E., & Blair, P. D. (2010). *Input-output analysis: Foundations and extensions*. Englewood Cliffs, NJ, USA: Prentice-Hall.
Murray, J., & Lenzen, M. (2013). *The practitioner's guide to multi-regional input-output analysis*. Champaign, USA: Common Ground.
Norton, M. I., & Ariely, D. (2011). Building a better America—One wealth quintile at a time. *Perspectives on Psychological Science, 6*(1), 9–12.
OECD. (2011a). *Divided we stand why inequality keeps rising*. Paris: OECD.
OECD. (2011b). *Growing income inequality in OECD countries: What drives it and how can policy tackle it?*. Paris: OECD.
Pavcnik, N., (2011). Globalization and within-country income inequality. In M. Bacchetta & M. Jansen (Eds.), *Making globalization socially sustainable* (pp. 233–259). Geneva: International Labour Organization and World Trade Organization.
Peters, G. P. (2010). Carbon footprints and embodied carbon at multiple scales. *Current Opinion in Environmental Sustainability, 2,* 245–250.
Solt, F. (2009). Standardizing the world income inequality database. *Social Science Quarterly, 90* (2), 231–242. SWIID Version 3.1, December 2011.
Tang, M. K. (2008). *Twin global shocks dent United Kingdom outlook*. European Department. Imf.org.
The Observatory of Economic Complexity. (2012). Internet site http://atlas.media.mit.edu/. USA: MIT.
UN. (2005). *The inequality predicament—Report on the world social situation 2005*. Economic and social affairs. New York: UN.
UN. (2009). *Rethinking poverty—Report on the world social situation 2010*. Department of Economic and Social Affairs. New York: UN.
UN. (2010). *Millennium development goals report 2010*. New York: UN.
UN Department of Economic and Social Affairs. (2010). *The inequality predicament: Report on the world social situation, 2005*. Social Policy and Development Division, United Nations.
UNCTAD. (2012). *Trade and development report, 2012*. Policies for inclusive and balanced growth. Geneva: United Nations.
UNSD. (2011). *UN comtrade—United Nations commodity trade statistics database*. Internet site comtrade.un.org/. New York, USA: United Nations Statistics Division (UNSD).
UN Systems Task Team. (2012). *Addressing inequalities: The heart of he post-2015 agenda and the future we want for all*. UN systems task team on the post-2015 UN development agenda. UN Systems Task Team.
United Nations Development Program. (2010). *The real wealth of nations: Pathways to human development. Human development report 2010*. New York: UNDP.
Verhoogen, E. A. (2008). Trade, quality upgrading, and wage inequality in the Mexican manufacturing sector. *The Quarterly Journal of Economics, 123*(2), 489–529.
Wiedmann, T., Wood, R., Lenzen, M., Minx, J., Guan, D., & Barrett, J. (2008). *Development of an embedded carbon emissions indicator—Producing a time series of input-output tables and embedded carbon dioxide emissions for the UK by using a MRIO data optimisation system*. University of York and University of Sydney, Stockholm Environment Institute; Centre for Integrated Sustainability Analysis. London, Department for Environment, Food and Rural Affairs.
World Bank. (2000). *Can Africa claim the 21st century?* Washington, DC: The Word Bank.

Chapter 10
Conclusion

In this work we have used a Multi-Regional Input-Output table coupled with satellite accounts that hold information about employment, income, occupational health and safety, poverty, unemployment, and others to calculate so called *Social Footprints of Nations*. Social footprints can play an important role in illustrating the impact of international trade on social indicators within countries. Thus, the results presented in this study are crucial in any policy formulation that targets human rights and working conditions in local, regional and global contexts. The conclusion beyond the results is that the agriculture sector is the key to improving working conditions through trade among all other sectors in an economy. This is because in most countries, especially developing ones, agriculture accounts for: high employment rates compared to other sectors; low wages with most workers living under the international poverty line; and a high number of accidents, fatal and non-fatal, in the workplace. In addition, there are many workers in the agriculture sector working purely to produce products consumed outside national borders. In spatial terms, African countries have the highest rate of embodied cases in exports for most of the investigated indicators.

It's impossible to isolate a crisis in one country from being spread into other neighbouring countries. The world is now more integrated and globalized, especially in trade. Hence, from that perspective, targeted policy to improve working conditions needs not only to be formulated within the local borders of producing countries, but also to include those in long supply chains (mostly developed countries) who benefit from those imported commodities. This study includes more than 95% of the world's workforce, about 150 countries and around 15 thousand sectors of the world economy.

The outcomes reported from this work can help to bring transparency for example, to corporate social responsibility, the fair trade movement and labour rights issues. The results are important to people working in international organizations (such as United Nations, Global Reporting Initiatives, World Bank and others) to inform them where their work should concentrated to narrow the distances to reach the implemented targets especially those in Millennium and Sustainable Development Goals. There are academics and people dealing with workers' rights who have published

A. Alsamawi et al., *The Social Footprints of Global Trade*,
Environmental Footprints and Eco-design of Products and Processes,
DOI 10.1007/978-981-10-4137-2_10

relevant research in the past years (Simas et al. 2014; Alsamawi et al. 2014; UNEP 2009). In these research projects, authors explore the hidden conditions of people working usually in the southern hemisphere to support the lifestyle of the northern hemisphere. Evidence of workplace conditions, such as the Rana Plaza building collapse in Bangladesh a few years ago, needs to alert us to do more for those workers. Environmentally, one of the solutions used successfully to reduce emissions and environmental degradation, was environmental labelling. Thus, labelling could also be used to guide and inform consumers to choose products that in addition to being environmentally-friendly, have benefits to humans in terms of wages, safety, and other conditions that relate to work. While the employment footprints show that about 20% of the world workforce is working in the export-oriented sector, social labelling could be a solution to lifting those workers, for instance, out of poverty through the international trade movement. Although consumers could have a huge influence on bringing about changes for those workers, corporations, governments, and NGOs can also play an important role to enhance and improve the whole reporting system. Developing a reporting system would help companies enhance their sustainability performance reporting, and in particular the reports that deal with social responsibility. The findings in this work show that countries that have high-embodied poverty in exports have also high-embodied OSH, employment, and inequality among all workers. While all these indicators are intertwined, improving one of them could have a great influence on the others.

This work could change our understanding not only of how increases in trade will help to lift people in poor countries out of poverty, but also of other indicators that need to be taken into account. Also, it could be a roadmap to NGOs to have a better understanding of how their work could have huge benefits from improving the challenging conditions of workers. Once again, the results reported in this work would be valuable for policymakers in formulating policies that target workers facing challenging workplaces.

The results presented in this work can be incorporated with other economic and environmental accounts to build a Triple Bottom Line inventory of the world economy.

References

Alsamawi, A., Murray, J., Lenzen, M., Moran, D., & Kanemoto, K. (2014). The inequality footprints of nations: A novel approach to quantitative accounting of income inequality. *PLoS ONE, 9*(10), e110881. doi:10.1371/journal.pone.0110881

Simas, M. S., Golsteijn, L., Huijbregts, M. A. J., Wood, R., & Hertwich, E. G. (2014). The "bad labor" footprint: quantifying the social impacts of globalization. *Sustain., 6,* 7514–7540. doi:10.3390/su6117514

UNEP (2009): *Guidelines for Social Like Cycle assessment of Products*, UN Environment Program, Life Cycle Initiative, Working Group on Social Impacts in LCA.

Appendix A
List of Countries Covered in This Paper

UN code	Name	Sectors (number of industries/products)
4	Afghanistan	26/0
8	Albania	26/0
12	Algeria	26/0
20	Andorra	26/0
24	Angola	26/0
28	Antigua and Barbuda	26/0
32	Argentina	125/196
51	Armenia	26/0
533	Aruba	26/0
36	Australia	345/345
40	Austria	61/61
31	Azerbaijan	26/0
44	Bahamas	26/0
48	Bahrain	26/0
50	Bangladesh	26/0
52	Barbados	26/0
112	Belarus	26/0
56	Belgium	61/61
84	Belize	26/0
204	Benin	26/0
60	Bermuda	26/0
64	Bhutan	26/0
68	Bolivia	37/37
70	Bosnia and Herzegovina	26/0
72	Botswana	26/0
76	Brazil	56/111
92	British Virgin Islands	26/0
96	Brunei Darussalam	26/0

(continued)

A. Alsamawi et al., *The Social Footprints of Global Trade*,
Environmental Footprints and Eco-design of Products and Processes,
DOI 10.1007/978-981-10-4137-2

(continued)

UN code	Name	Sectors (number of industries/products)
100	Bulgaria	26/0
854	Burkina Faso	26/0
108	Burundi	26/0
116	Cambodia	26/0
120	Cameroon	26/0
124	Canada	49/0
132	Cape Verde	26/0
136	Cayman Islands	26/0
140	Central African Republic	26/0
148	Chad	26/0
152	Chile	75/75
156	China	0/123
170	Colombia	60/60
178	Congo	26/0
188	Costa Rica	26/0
191	Croatia	26/0
192	Cuba	26/0
196	Cyprus	26/0
203	Czech Republic	61/61
384	Côte d'Ivoire	26/0
408	Democratic People's Republic of Korea	26/0
180	Democratic Republic of the Congo, previously Zaïre	26/0
208	Denmark	131/0
262	Djibouti	26/0
214	Dominican Republic	26/0
218	Ecuador	49/61
818	Egypt	26/0
222	El Salvador	26/0
232	Eritrea	26/0
233	Estonia	61/61
231	Ethiopia	26/0
242	Fiji	26/0
246	Finland	61/61
250	France	61/61
258	French Polynesia	26/0
266	Gabon	26/0
270	Gambia	26/0
268	Georgia	47/68

(continued)

(continued)

UN code	Name	Sectors (number of industries/products)
276	Germany	0/72
288	Ghana	26/0
300	Greece	61/61
304	Greenland	31/0
320	Guatemala	26/0
324	Guinea	26/0
328	Guyana	26/0
332	Haiti	26/0
340	Honduras	26/0
344	Hong Kong	38/38
348	Hungary	61/61
352	Iceland	26/0
356	India	116/116
360	Indonesia	0/77
364	Iran	100/148
368	Iraq	26/0
372	Ireland	61/61
376	Israel	163/163
380	Italy	61/61
388	Jamaica	26/0
392	Japan	0/402
400	Jordan	26/0
398	Kazakhstan	0/121
404	Kenya	51/51
414	Kuwait	55/0
417	Kyrgyzstan	89/87
418	Lao People's Democratic Republic	26/0
428	Latvia	61/61
422	Lebanon	26/0
426	Lesotho	26/0
430	Liberia	26/0
434	Libyan Arab Jamahiriya	26/0
438	Liechtenstein	26/0
440	Lithuania	61/61
442	Luxembourg	26/0
446	Macao Special Administrative Region of China	26/0
450	Madagascar	26/0
454	Malawi	26/0

(continued)

(continued)

UN code	Name	Sectors (number of industries/products)
458	Malaysia	0/98
462	Maldives	26/0
466	Mali	26/0
470	Malta	61/61
478	Mauritania	26/0
480	Mauritius	57/67
484	Mexico	80/80
492	Monaco	26/0
496	Mongolia	26/0
499	Montenegro	26/0
504	Morocco	26/0
508	Mozambique	26/0
104	Myanmar	26/0
516	Namibia	26/0
524	Nepal	26/0
528	Netherlands	61/61
530	Netherlands Antilles	16/41
540	New Caledonia	26/0
554	New Zealand	127/210
558	Nicaragua	26/0
562	Niger	26/0
566	Nigeria	26/0
578	Norway	61/61
275	West Bank and Gaza Strip	26/0
512	Oman	26/0
586	Pakistan	26/0
591	Panama	26/0
598	Papua New Guinea	26/0
600	Paraguay	34/47
604	Peru	46/46
608	Philippines	0/77
616	Poland	61/61
620	Portugal	61/61
634	Qatar	26/0
410	Republic of Korea	0/78
498	Republic of Moldova	26/0
642	Romania	61/61
643	Russian Federation	49/0
646	Rwanda	26/0

(continued)

(continued)

UN code	Name	Sectors (number of industries/products)
882	Samoa	26/0
674	San Marino	26/0
678	Sao Tome and Principe	26/0
682	Saudi Arabia	26/0
686	Senegal	26/0
688	Serbia	26/0
690	Seychelles	26/0
694	Sierra Leone	26/0
702	Singapore	154/154
703	Slovakia	61/61
705	Slovenia	61/61
706	Somalia	26/0
710	South Africa	95/96
724	Spain	76/119
144	Sri Lanka	26/0
736	Sudan	26/0
740	Suriname	26/0
748	Swaziland	26/0
752	Sweden	61/61
756	Switzerland	43/43
760	Syrian Arab Republic	26/0
761	Taiwan	0/163
762	Tajikistan	26/0
764	Thailand	0/180
807	Macedonia	61/61
768	Togo	26/0
780	Trinidad and Tobago	26/0
788	Tunisia	26/0
792	Turkey	61/61
795	Turkmenistan	26/0
800	Uganda	26/0
804	Ukraine	0/121
784	United Arab Emirates	26/0
826	United Kingdom	511/511
834	Tanzania	26/0
840	United States	429/429
858	Uruguay	84/103
860	Uzbekistan	0/123
548	Vanuatu	26/0

(continued)

(continued)

UN code	Name	Sectors (number of industries/products)
862	Venezuela	122/122
704	Viet Nam	0/113
887	Yemen	26/0
894	Zambia	26/0
716	Zimbabwe	26/0

Appendix B
Countries and Years for Which Data Are Available

Country	1990	1991	1992	1993	1994	1995	1996	1997	1998	1999	2000
Afghanistan	0										
Albania	0						x	x	x	x	x
Algeria	x	x	x	x	x	x	x	x	x	x	x
Andorra	0										
Angola	0					x	x	x	x	x	x
Antigua and Barbuda	0										
Argentina	x	x	x	x	x	x	x	x	x	x	x
Armenia	x	x	x	x	x	x	x	x	x	x	x
Aruba	0										
Australia	x	x	x	x	x	x	x	x	x	x	x
Austria	x	x	x	x	x	x	x	x	x	x	x
Azerbaijan	x	x	x	x	x	x	x	x	x	x	x
Bahamas	x	x	x	x	x	x	x	x	x	x	x
Bahrain	0										
Bangladesh	x	x	x	x	x	x	x	x	x	x	x
Barbados	x	x	x	x	x	x	x	x			
Belarus	x	x	x	x	x	x	x	x	x	x	x
Belgium	x	x	x	x	x	x	x	x	x	x	x
Belize	0			x	x	x	x	x	x	x	
Benin	0										
Bermuda	0										
Bhutan	0										
Bolivia	x	x	x	x	x	x	x	x	x	x	x
Bosnia and Herzegovina	0	x	x	x	x	x	x	x	x	x	x
Botswana	x	x	x	x	x	x	x	x	x	x	x
Brazil	x	x	x	x	x	x	x	x	x	x	x

(continued)

A. Alsamawi et al., *The Social Footprints of Global Trade*,
Environmental Footprints and Eco-design of Products and Processes,
DOI 10.1007/978-981-10-4137-2

(continued)

Country	1990	1991	1992	1993	1994	1995	1996	1997	1998	1999	2000
British Virgin Islands	0										
Brunei Darussalam	0										
Bulgaria	x	x	x	x	x	x	x	x	x	x	x
Burkina Faso	0				x	x	x	x	x	x	x
Burundi	0		x	x	x	x	x	x	x	x	x
Cambodia	0				x	x	x	x	x	x	x
Cameroon	x	x	x	x	x	x	x	x	x	x	x
Canada	x	x	x	x	x	x	x	x	x	x	x
Cape Verde	x	x	x	x	x	x	x	x	x	x	x
Cayman Islands	0		x								
Central African Republic			x	x	x	x	x	x	x	x	x
Chad											
Chile	x	x	x	x	x	x	x	x	x	x	x
China	x	x	x	x	x	x	x	x	x	x	x
Colombia	x	x	x	x	x	x	x	x	x	x	x
Congo	0										
Costa Rica	x	x	x	x	x	x	x	x	x	x	x
Croatia	x	x	x	x	x	x	x	x	x	x	x
Cuba	0										
Cyprus	x	x	x	x	x	x	x	x	x	x	x
Czech Republic	x	x	x	x	x	x	x	x	x	x	x
Côte d'ivoire	x	x	x	x	x	x	x	x	x	x	x
North Korea	x						x		x		
Democratic Republic of the Congo, previously Zaïre	0										
Denmark	x	x	x	x	x	x	x	x	x	x	x
Djibouti						x	x	x	x	x	x
Dominican Republic	x	x	x	x	x	x	x	x	x	x	x
Ecuador	x	x	x	x	x	x	x	x	x	x	x
Egypt	x	x	x	x	x	x	x	x	x	x	x
El Salvador	x	x	x	x	x	x	x	x	x	x	x
Eritrea	0										

(continued)

(continued)

Country	1990	1991	1992	1993	1994	1995	1996	1997	1998	1999	2000
Estonia	x	x	x	x	x	x	x	x	x	x	x
Ethiopia	x	x	x	x	x	x	x	x	x	x	x
Fiji	x	x	x						x		x
Finland	x	x	x	x	x	x	x	x	x	x	x
France	x	x	x	x	x	x	x	x	x	x	x
French Polynesia	0										
Gabon	x	x	x								
Gambia	x	x	x	x	x	x	x	x	x	x	x
Georgia	x	x	x	x	x	x	x	x	x	x	x
Germany	x	x	x	x	x	x	x	x	x	x	x
Ghana	x	x	x	x	x	x	x	x	x	x	x
Greece	x	x	x	x	x	x	x	x	x	x	x
Greenland	0				x						
Guatemala	x	x	x	x	x	x	x	x	x	x	x
Guinea	x	x	x	x	x	x	x	x	x	x	x
Guyana			x	x	x	x	x	x	x	x	
Haiti	x	x	x	x	x	x	x	x	x	x	x
Honduras	x	x	x	x	x	x	x	x	x	x	x
Hong Kong	x	x	x	x	x	x	x	x	x	x	x
Hungary	x	x	x	x	x	x	x	x	x	x	x
Iceland	0	x	x	x	x	x	x	x	x	x	x
India	x	x	x	x	x	x	x	x	x	x	x
Indonesia	x	x	x	x	x	x	x	x	x	x	x
Iran	x	x	x	x	x	x	x	x	x	x	x
Iraq											
Ireland	x	x	x	x	x	x	x	x	x	x	x
Israel	x	x	x	x	x	x	x	x	x	x	x
Italy	x	x	x	x	x	x	x	x	x	x	x
Jamaica	x	x	x	x	x	x	x	x	x	x	x
Japan	x	x	x	x	x	x	x	x	x	x	x
Jordan	x	x	x	x	x	x	x	x	x	x	x
Kazakhstan	x	x	x	x	x	x	x	x	x	x	x
Kenya	x	x	x	x	x	x	x	x	x	x	x
Kuwait	0										
Kyrgyzstan	x	x	x	x	x	x	x	x	x	x	x
Lao People's Democratic Republic	0		x	x	x	x	x	x	x	x	x
Latvia	x	x	x	x	x	x	x	x	x	x	x
Lebanon								x	x	x	x

(continued)

(continued)

Country	1990	1991	1992	1993	1994	1995	1996	1997	1998	1999	2000
Lesotho	x	x	x	x	x	x	x	x	x	x	x
Liberia	x										
Libyan Arab Jamahiriya	0										
Liechtenstein	0										
Lithuania	x	x	x	x	x	x	x	x	x	x	x
Luxembourg	x	x	x	x	x	x	x	x	x	x	x
Macau	x	x	x	x	x	x	x	x	x	x	x
Madagascar	x	x	x	x	x	x	x	x	x	x	x
Malawi	x	x	x	x	x	x	x	x	x	x	x
Malaysia	x	x	x	x	x	x	x	x	x	x	x
Maldives	0								x		
Mali	x	x	x	x	x	x	x	x	x	x	x
Malta	0										x
Mauritania	x	x	x	x	x	x	x	x	x	x	x
Mauritius	x	x	x	x	x	x	x	x	x	x	x
Mexico	x	x	x	x	x	x	x	x	x	x	x
Monaco	0		x					x	x	x	
Mongolia	0					x	x	x	x	x	x
Montenegro	0										x
Morocco	x	x	x	x	x	x	x	x	x	x	x
Mozambique	0						x	x	x	x	x
Myanmar	0										
Namibia				x	x	x	x	x	x	x	x
Nepal	x	x	x	x	x	x	x	x	x	x	x
Netherlands	x	x	x	x	x	x	x	x	x	x	x
Netherlands Antilles	0										
New Caledonia	0										
New Zealand	x	x	x	x	x	x	x	x	x	x	x
Nicaragua	x			x					x		
Niger		x	x	x	x	x	x	x	x	x	x
Nigeria	x	x	x	x	x	x	x	x	x	x	x
Norway	x	x	x	x	x	x	x	x	x	x	x
West Bank and Gaza Strip	0										
Oman	0										
Pakistan	x	x	x	x	x	x	x	x	x	x	x
Panama	x	x	x	x	x	x	x	x	x	x	x

(continued)

(continued)

Country	1990	1991	1992	1993	1994	1995	1996	1997	1998	1999	2000
Papua New Guinea	0					x	x	x	x	x	x
Paraguay	x	x	x	x	x	x	x	x	x	x	x
Peru	x	x	x	x	x	x	x	x	x	x	x
Philippines	x	x	x	x	x	x	x	x	x	x	x
Poland	x	x	x	x	x	x	x	x	x	x	x
Portugal	x	x	x	x	x	x	x	x	x	x	x
Qatar	0										
Republic of Korea	x	x	x	x	x	x	x	x	x	x	x
Moldova	x	x	x	x	x	x	x	x	x	x	x
Romania	x	x	x	x	x	x	x	x	x	x	x
Russian Federation	x	x	x	x	x	x	x	x	x	x	x
Rwanda	0					x	x	x	x	x	x
Samoa	0								x		
San Marino	0										
Sao Tome and Principe	0										
Saudi Arabia	0										
Senegal	0	x	x	x	x	x	x	x	x	x	x
Serbia											
Seychelles	0										x
Sierra Leone	x	x	x	x	x	x	x	x	x	x	x
Singapore	x	x	x	x	x	x	x	x	x	x	x
Slovakia	x	x	x	x	x	x	x	x	x	x	x
Slovenia	x	x	x	x	x	x	x	x	x	x	x
Somalia	0										
South Africa	x	x	x	x	x	x	x	x	x	x	x
Spain	x	x	x	x	x	x	x	x	x	x	x
Sri Lanka	x	x	x	x	x	x	x	x	x	x	x
Sudan	0										
Suriname	0									x	x
Swaziland	x				x	x	x	x	x	x	x
Sweden	x	x	x	x	x	x	x	x	x	x	x
Switzerland	x	x	x	x	x	x	x	x	x	x	x
Syrian Arab Republic	0										
Taiwan	x	x	x	x	x	x	x	x	x	x	x
Tajikistan	x	x	x	x	x	x	x	x	x	x	x
Thailand	x	x	x	x	x	x	x	x	x	x	x

(continued)

(continued)

Country	1990	1991	1992	1993	1994	1995	1996	1997	1998	1999	2000
Macedonia	0								x		x
Togo											
Trinidad and Tobago	x	x	x	x	x	x	x	x	x	x	x
Tunisia	x	x	x	x	x	x	x	x	x	x	x
Turkey	x	x	x	x	x	x	x	x	x	x	x
Turkmenistan	x	x	x	x	x	x	x	x	x	x	x
Uganda	x	x	x	x	x	x	x	x	x	x	x
Ukraine	x	x	x	x	x	x	x	x	x	x	x
United Arab Emirates	0										
United Kingdom	x	x	x	x	x	x	x	x	x	x	x
Tanzania	x	x	x	x	x	x	x	x	x	x	x
United States	x	x	x	x	x	x	x	x	x	x	x
Uruguay	x	x	x	x	x	x	x	x	x	x	x
Uzbekistan	x	x	x	x	x	x	x	x	x	x	x
Vanuatu	0				x						
Venezuela	x	x	x	x	x	x	x	x	x	x	x
Viet Nam	x	x	x	x	x	x	x	x	x	x	x
Yemen			x	x	x	x	x	x	x	x	x
Zambia	x	x	x	x	x	x	x	x	x	x	x
Zimbabwe	x	x	x	x	x	x					

Country	2001	2002	2003	2004	2005	2006	2007	2008	2009	2010
Afghanistan								x		
Albania	x	x	x	x	x			x		
Algeria	x	x	x	x	x					
Andorra	x	x	x					x		
Angola	x	x	x	x	x					x
Antigua and Barbuda	x	x								
Argentina	x	x	x	x	x	x	x	x	x	x
Armenia	x	x	x	x	x	x	x			
Aruba						x				
Australia	x	x	x	x	x	x	x	x	x	x
Austria	x	x	x	x	x	x	x	x	x	x
Azerbaijan	x	x	x	x	x	x	x	x		
Bahamas	x	x	x	x						
Bahrain				x						
Bangladesh	x	x	x	x	x	x	x	x	x	x
Barbados										
Belarus	x	x	x	x	x	x	x	x		

(continued)

(continued)

Country	2001	2002	2003	2004	2005	2006	2007	2008	2009	2010
Belgium	x	x	x	x	x	x	x	x	x	x
Belize					x	x				
Benin			x	x	x	x				
Bermuda						x				
Bhutan			x	x	x	x	x			
Bolivia	x	x	x	x	x	x	x			
Bosnia and Herzegovina	x	x	x	x	x		x		x	
Botswana	x	x	x	x	x				x	x
Brazil	x	x	x	x	x	x	x	x	x	x
British Virgin Islands					x					
Brunei Darussalam	x									
Bulgaria	x	x	x	x	x	x	x	x	x	x
Burkina Faso	x	x	x						x	
Burundi	x	x	x	x	x	x				
Cambodia	x	x	x	x			x	x		x
Cameroon	x	x					x			
Canada	x	x	x	x	x	x	x	x	x	
Cape Verde	x	x	x	x	x					
Cayman Islands			x					x		
Central African Republic	x	x	x					x		
Chad		x	x	x	x					
Chile	x	x	x	x	x	x	x	x	x	
China	x	x	x	x	x	x			x	x
Colombia	x	x	x	x	x	x	x	x	x	
Congo					x	x				
Costa Rica	x	x	x	x	x	x	x	x	x	
Croatia	x	x	x	x	x	x	x	x	x	
Cuba		x								
Cyprus	x	x	x	x	x	x	x	x	x	
Czech Republic	x	x	x	x	x	x	x	x	x	x
Côte d'ivoire	x	x								
North Korea					x					x
Democratic Republic of the Congo, previously Zaïre				x						
Denmark	x	x	x	x	x	x	x	x	x	x
Djibouti	x	x	x	x	x					
Dominican Republic	x	x	x	x	x	x	x	x	x	x
Ecuador	x	x	x	x	x	x	x	x	x	

(continued)

(continued)

Country	2001	2002	2003	2004	2005	2006	2007	2008	2009	2010
Egypt	x	x	x	x	x	x	x	x		
El Salvador	x	x	x	x	x	x	x	x		
Eritrea								x		
Estonia	x	x	x	x	x	x	x	x	x	x
Ethiopia	x	x	x	x	x					
Fiji			x						x	
Finland	x	x	x	x	x	x	x	x	x	x
France	x	x	x	x	x	x	x	x	x	x
French Polynesia		x								
Gabon					x					
Gambia	x	x	x							
Georgia	x	x	x	x	x	x				
Germany	x	x	x	x	x	x	x	x	x	x
Ghana	x	x	x	x	x	x				
Greece	x	x	x	x	x	x	x	x	x	x
Greenland					x					
Guatemala	x	x	x	x	x	x				
Guinea	x	x	x	x	x	x				
Guyana				x						
Haiti	x					x	x	x		
Honduras	x	x	x	x	x	x	x	x	x	
Hong Kong	x	x	x	x	x	x		x		
Hungary	x	x	x	x	x	x	x	x	x	x
Iceland	x	x	x	x	x	x	x	x	x	x
India	x	x	x	x	x					
Indonesia	x	x	x	x	x	x	x	x	x	x
Iran	x	x	x	x						
Iraq			x	x			x			
Ireland	x	x	x	x	x	x	x	x	x	
Israel	x	x	x	x	x				x	x
Italy	x	x	x	x	x	x	x	x	x	x
Jamaica	x	x	x	x						
Japan	x	x	x	x	x	x	x	x	x	x
Jordan	x	x	x	x	x	x		x		
Kazakhstan	x	x	x	x	x	x	x	x	x	x
Kenya	x	x	x	x	x					
Kuwait					x					
Kyrgyzstan	x	x	x	x	x	x	x	x	x	x
Lao People's Democratic Republic	x	x	x	x	x	x	x			

(continued)

(continued)

Country	2001	2002	2003	2004	2005	2006	2007	2008	2009	2010
Latvia	x	x	x	x	x	x	x	x	x	x
Lebanon	x	x	x	x	x				x	
Lesotho	x	x	x	x	x					
Liberia					x	x	x			
Libyan Arab Jamahiriya	x									
Liechtenstein		x								
Lithuania	x	x	x	x	x	x	x	x	x	x
Luxembourg	x	x	x	x	x	x	x	x	x	x
Macau	x	x	x	x	x	x	x			
Madagascar	x	x	x	x	x					
Malawi	x	x	x	x	x					
Malaysia	x	x	x	x	x		x		x	
Maldives			x	x						
Mali	x	x	x	x	x	x				
Malta	x	x	x	x	x	x	x	x	x	x
Mauritania				x				x		
Mauritius	x	x	x	x	x	x				
Mexico	x	x	x	x	x	x	x	x	x	x
Monaco	x	x	x	x	x	x	x	x	x	x
Mongolia	x	x	x	x	x	x				
Montenegro	x	x	x	x	x	x	x	x		
Morocco	x	x	x	x	x	x	x			
Mozambique	x	x	x	x	x					
Myanmar				x						
Namibia	x	x	x	x	x			x		
Nepal	x	x	x	x						
Netherlands	x	x	x	x	x	x	x	x	x	x
Netherlands Antilles	x									
New Caledonia					x					
New Zealand	x	x	x	x	x	x	x			
Nicaragua	x				x					
Niger	x	x	x	x	x					
Nigeria	x	x	x	x						
Norway	x	x	x	x	x	x	x	x	x	x
West Bank and Gaza Strip							x		x	
Oman	x									
Pakistan	x	x	x	x	x	x	x			
Panama	x	x	x	x	x	x	x	x	x	

(continued)

(continued)

Country	2001	2002	2003	2004	2005	2006	2007	2008	2009	2010
Papua New Guinea	x	x	x	x	x					
Paraguay	x	x	x	x	x	x	x	x	x	
Peru	x	x	x	x	x	x	x	x	x	
Philippines	x	x	x	x	x	x	x	x	x	
Poland	x	x	x	x	x	x	x	x	x	x
Portugal	x	x	x	x	x	x	x	x	x	x
Qatar							x			
Republic of Korea	x	x	x	x	x	x	x	x	x	x
Moldova	x	x	x	x	x	x	x	x	x	x
Romania	x	x	x	x	x	x	x	x	x	x
Russian Federation	x	x	x	x	x	x	x	x	x	
Rwanda	x	x	x	x	x	x				
Samoa										
San Marino				x						
Sao Tome and Principe	x									
Saudi Arabia			x							
Senegal	x	x	x	x	x					
Serbia	x	x	x	x	x	x	x			
Seychelles							x	x		
Sierra Leone	x	x	x	x	x					
Singapore	x	x	x	x	x	x	x	x	x	
Slovakia	x	x	x	x	x	x	x	x	x	x
Slovenia	x	x	x	x	x	x	x	x	x	x
Somalia		x								
South Africa	x	x	x	x	x	x	x	x	x	
Spain	x	x	x	x	x	x	x	x	x	x
Sri Lanka	x	x					x			
Sudan	x									
Suriname	x	x	x	x	x					
Swaziland	x	x	x	x	x				x	x
Sweden	x	x	x	x	x	x	x	x	x	x
Switzerland	x	x	x	x	x	x	x	x	x	
Syrian Arab Republic			x							
Taiwan	x	x	x	x	x					
Tajikistan	x	x	x	x						
Thailand	x	x	x	x		x		x	x	
Macedonia		x	x	x	x	x		x	x	x
Togo					x	x				
Trinidad and Tobago	x	x	x	x	x			x	x	x
Tunisia	x	x	x	x	x					

(continued)

(continued)

Country	2001	2002	2003	2004	2005	2006	2007	2008	2009	2010
Turkey	x	x	x	x	x	x	x	x	x	
Turkmenistan	x	x	x	x	x					
Uganda	x	x	x	x	x	x				
Ukraine	x	x	x	x	x	x	x			
United Arab Emirates					x					
United Kingdom	x	x	x	x	x	x	x	x	x	x
Tanzania	x						x			
United States	x	x	x	x	x	x	x	x	x	x
Uruguay	x	x	x	x	x	x	x	x	x	
Uzbekistan	x	x	x	x	x					
Vanuatu								x		
Venezuela	x	x	x	x	x	x	x	x	x	x
Viet Nam	x	x	x	x	x	x				
Yemen	x	x	x	x	x					
Zambia	x	x	x	x	x					
Zimbabwe										

x data available

Data sources [SWIID version 3.1 Solt Frederick (2009)], OCED database, and the World Bank database

Appendix C

C.1 What Is a Satellite Account?

Satellite accounts are used to expand the analytical capacity of national economic accounts in social and environmental areas (UNSD 2003). Moreover, they link physical data sources to the monetary national account (UNSD 2009). Satellite accounts are expressed in terms of satellite indicators.

Satellite indicators represent social and environmental impacts. For example, employment is an indicator of a particular social impact. Most satellite indicators can be broken down into sub-indicators (Costantini et al. 2009) (Tables C.1 and C.2).

There are two well-known environmental account systems:

UN System of Environmental-Economic Accounting SEEA: A satellite account of the System of National Accounts (SNA) that supports decision-making (UNSD 2003). The SEEA is of benefit to academic researchers, scientists, and agencies providing information that can be used to underpin their decisions (Bartelmus 2007; Dietz and Neumayer 2007; Lange 2007; Lenzen and Murray 2011; Smith 2006; Walker and Pearson 2007). In this work, we use the UN SEEA system for compiling an inequality satellite account.

NAMEA: In the late 1980s, the Central Bureau of Statistics of the Netherlands developed a system for describing environmental flows in conjunction with the National Accounts. The system, known as the National Accounting Matrix including Environmental Accounts (NAMEA) creates a link between the National Accounts and environmental statistics (de Haan and Keuning 1996) consumption and production patterns (EC 2001; Ike 1999; Keuning et al. 1999; Vaze 1999).

C.2 Basic Definitions of Social Indicators

Horn (1993) describes social indicators as statistical measures of non-monetary factors that mainly deal with human performance. Gallopin (1997) defines the social indicator as a variable that is usually used to describe an attribute of a phenomenon. The OECD (2009) describes social indicators as a mirrors that reflect

A. Alsamawi et al., *The Social Footprints of Global Trade*,
Environmental Footprints and Eco-design of Products and Processes,
DOI 10.1007/978-981-10-4137-2

Table C.1 Example of social indicators

Indicators	Sub-indicators
Job/employment generation	Male
	Female
	Child labour
Family income	Male
	Female
	Age group

Table C.2 Example of environmental indicators

Indicators	Sub-indicators
Greenhouse gas emissions	CO_2
	CH_4
	N_2O
	HFC
Energy use	Black coal
	Electricity
	Natural gas
	Auto gasoline, leaded
Water use	Mains water
	Self-supplied water
	Reuse water
	In-stream water

most of our social activities such as health and education. In this regard, the social indicator may be considered as an intermediary that describes the phenomenon.

C.3 Constructing the Satellite and Income Satellite Accounts

This section is kept brief because the construction principles of global satellite accounts have been determined elsewhere (Lenzen et al. 2012).

There are three important steps needed to construct a satellite account: identify databases; prepare constraints data; and create a concordance matrix.

1. *Identify databases*

There are many data sources that provide information about different indicators (such as http://data.un.org/). These indicators may have been collected at different times and locations. Sometimes, they are broken down into sub-indices. These sub-indices usually provide more information in a complex system, for example, GHG (indicator) has CH_4 as sub-indicator, and CH_4 also has fuel as sub-index. In some cases, further disaggregation is needed to provide a deeper understanding of the indicator. These hierarchical levels are meaningful and useful.

2. *Prepare constraints data*

Input-Output tables are often subject to constraints. A constraint is a mathematical condition imposed onto elements of global Multi-Region Input-Output (MRIO) tables. These constraints can be expressed in mathematical equations (Lenzen et al. 2012). Constraints data need to be prepared with names that distinguishes and saves the data according to specific parameters. These parameters are separated by underscore or slash characters to fit the global MRIO system and the output is saved as comma separated files (csv).

3. *Concordances*

We used employment data from the International Labour Organization database (laborsta). These databases are classified by International Standard Industrial Classification ISIC versions (1, 2, 3, and 4). To link between these different versions we prepared a concordance matrix. To reach a high level of disaggregation, we linked the classifications to the Harmonized Commodity Description and Coding System (HS), which is, comprises about 5000 commodity groups (see Lenzen et al. 2012).

Appendix D
Further Comparisons of Lorenz Curves Based on Measured Income and Employment Distributions and Fitted Based on Gini Index Data

Below are some examples of Lorenz curves calculated from income and employment quintiles or deciles to determine the Gini index of nations [circles represent the data (ABS 2004, 2010; Family Spending 2010, 2011; IBGE 2010) and the lines represent the power function approximation we used (these data populate *I* and *P* in Sect. C.4)].

A. Alsamawi et al., *The Social Footprints of Global Trade*,
Environmental Footprints and Eco-design of Products and Processes,
DOI 10.1007/978-981-10-4137-2

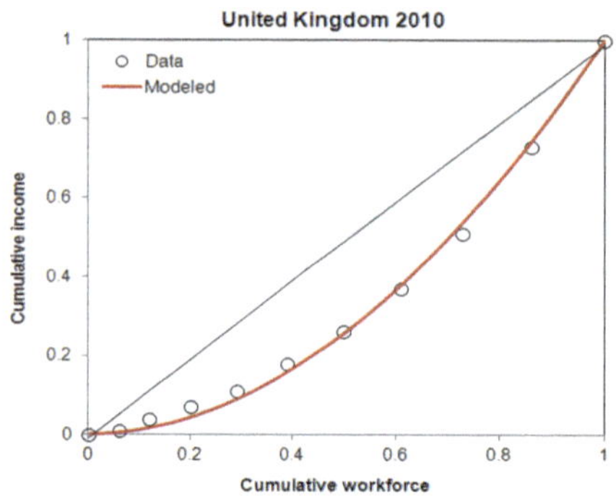
United Kingdom 2010
Data
Modeled
Cumulative income
Cumulative workforce

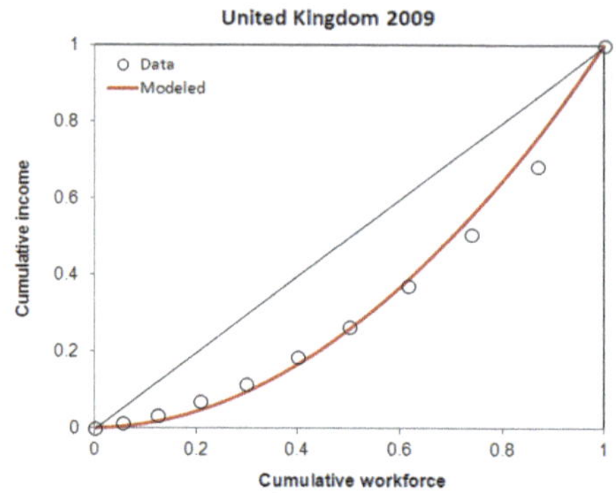
United Kingdom 2009
Data
Modeled
Cumulative income
Cumulative workforce

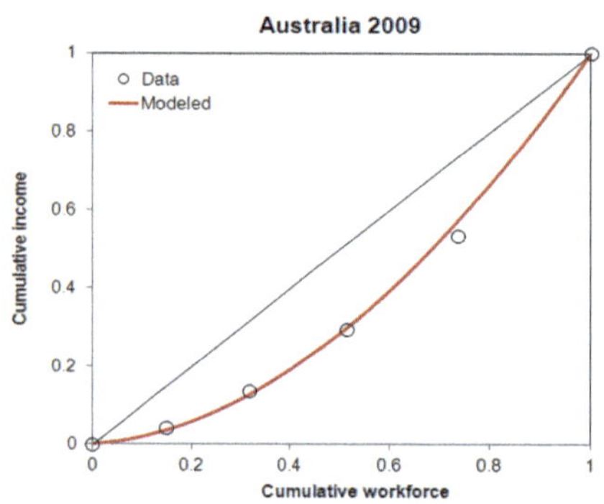
Australia 2009
Data
Modeled
Cumulative income
Cumulative workforce

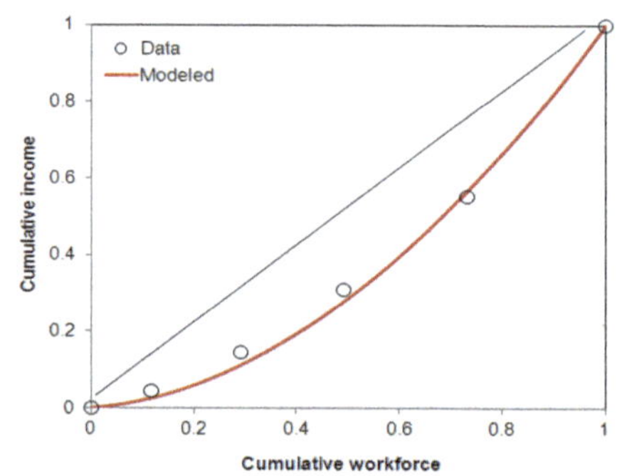
Data
Modeled
Cumulative income
Cumulative workforce

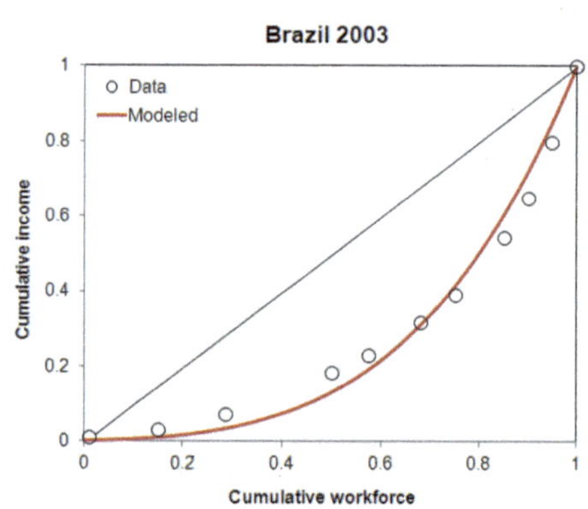
Brazil 2003
Data
Modeled
Cumulative income
Cumulative workforce

Appendix E

Table E.1 The ranking of added countries

Added-countries	Rank
France	35
Republic of Korea	40
United Kingdom	55
United States	63
Thailand	141
China	154

This table shows the ranking of the countries that were added to Fig. 3. Our research covered 187 countries. They were added because of their significant share of the world's Gross Domestic Product (GDP).

Additional information that supports Fig. 3 and Table 1 in the manuscript.

Country	$\frac{\text{Inequality footprint}}{\text{Gini index of domestic workforce}}$	Inequality footprint	Within-country Gini index
Norway	1.31	0.31	0.24
Slovenia	1.26	0.30	0.24
Hungary	1.26	0.30	0.24
Netherlands	1.25	0.32	0.26
Finland	1.25	0.32	0.25
Iceland	1.23	0.32	0.26
Sweden	1.23	0.30	0.24
Japan	1.19	0.34	0.29
Canada	1.14	0.34	0.30
Austria	1.14	0.30	0.26
Switzerland	1.13	0.33	0.30
Denmark	1.12	0.30	0.27
Australia	1.09	0.34	0.31
Germany	1.09	0.32	0.29

(continued)

A. Alsamawi et al., *The Social Footprints of Global Trade*,
Environmental Footprints and Eco-design of Products and Processes,
DOI 10.1007/978-981-10-4137-2

(continued)

Country	$\frac{\text{Inequality footprint}}{\text{Gini index of domestic workforce}}$	Inequality footprint	Within-country Gini index
France	1.06	0.32	0.30
Republic of Korea	1.04	0.33	0.32
United Kingdom	0.97	0.32	0.33
United States	0.95	0.34	0.36
Thailand	0.78	0.31	0.40
China	0.75	0.32	0.42
Philippines	0.75	0.32	0.43
Mexico	0.72	0.35	0.48
Malaysia	0.70	0.33	0.46
Brazil	0.64	0.35	0.55
Russian Federation	0.57	0.27	0.46
South Africa	0.53	0.34	0.63

This table supports the detail of Fig. 3 and Table 1, it contains: ratio of inequality footprint to within-country Gini index, inequality footprint (populate $\overline{\gamma}_X$ in Sect. C.5), and within-country Gini index (populate γ in Sect. C.4).

Appendix F

Example of time series of within-country Gini index (populate γ in Sect. C.4) against inequality footprint (populate $\overline{\gamma}_X$ in Sect. C.5) spanning 1990–2010 for the top and bottom two countries of Fig. 3 and including United Kingdom and United States (each year is represented by a dot point and 2010 is represented by the arrowhead) (see Tables F.1, F.2 and F.3).

A. Alsamawi et al., *The Social Footprints of Global Trade*,
Environmental Footprints and Eco-design of Products and Processes,
DOI 10.1007/978-981-10-4137-2

Table F.1 Additional information that supports Fig. F.1.

Year	Within country Gini index	Inequality footprint	Within country Gini index	Inequality footprint	Within country Gini index	Inequality footprint	Within country Gini index	Inequality footprint	Within country Gini index	Inequality footprint	Within country Gini index	Inequality footprint
	Norway	Norway	Slovenia	Slovenia	United States	United States	Russian Federation	Russian Federation	South Africa	South Africa	United Kingdom	United Kingdom
1990	0.23	0.31	0.26	0.31	0.34	0.33	0.24	0.25	0.62	0.34	0.32	0.31
1991	0.23	0.31	0.26	0.31	0.34	0.33	0.33	0.25	0.62	0.35	0.32	0.32
1992	0.23	0.31	0.26	0.30	0.34	0.33	0.40	0.24	0.60	0.35	0.33	0.32
1993	0.23	0.31	0.28	0.30	0.35	0.32	0.40	0.26	0.59	0.34	0.33	0.31
1994	0.24	0.30	0.28	0.30	0.36	0.32	0.42	0.28	0.58	0.33	0.34	0.31
1995	0.24	0.30	0.28	0.30	0.36	0.32	0.45	0.31	0.58	0.33	0.34	0.31
1996	0.23	0.30	0.28	0.30	0.37	0.32	0.42	0.29	0.60	0.33	0.35	0.31
1997	0.23	0.31	0.28	0.30	0.37	0.33	0.40	0.29	0.58	0.34	0.35	0.31
1998	0.23	0.31	0.28	0.30	0.37	0.34	0.40	0.29	0.60	0.34	0.36	0.32
1999	0.24	0.31	0.27	0.30	0.37	0.34	0.41	0.27	0.62	0.33	0.36	0.32
2000	0.25	0.31	0.25	0.31	0.37	0.34	0.43	0.28	0.65	0.33	0.36	0.32
2001	0.26	0.32	0.24	0.31	0.37	0.35	0.42	0.29	0.65	0.34	0.35	0.33
2002	0.26	0.32	0.22	0.31	0.37	0.36	0.42	0.28	0.66	0.34	0.35	0.33
2003	0.27	0.30	0.22	0.29	0.37	0.34	0.43	0.27	0.67	0.33	0.34	0.31
2004	0.25	0.30	0.23	0.29	0.37	0.34	0.44	0.26	0.68	0.33	0.34	0.31
2005	0.28	0.30	0.24	0.29	0.37	0.34	0.45	0.26	0.68	0.33	0.35	0.31
2006	0.31	0.31	0.24	0.30	0.37	0.35	0.46	0.27	0.67	0.34	0.33	0.32
2007	0.24	0.32	0.23	0.31	0.37	0.35	0.46	0.28	0.66	0.34	0.33	0.32
2008	0.25	0.32	0.23	0.30	0.36	0.34	0.46	0.26	0.65	0.34	0.34	0.32
2009	0.24	0.31	0.23	0.30	0.36	0.34	0.46	0.27	0.64	0.34	0.32	0.32
2010	0.24	0.31	0.24	0.30	0.36	0.34	0.46	0.27	0.63	0.34	0.33	0.32

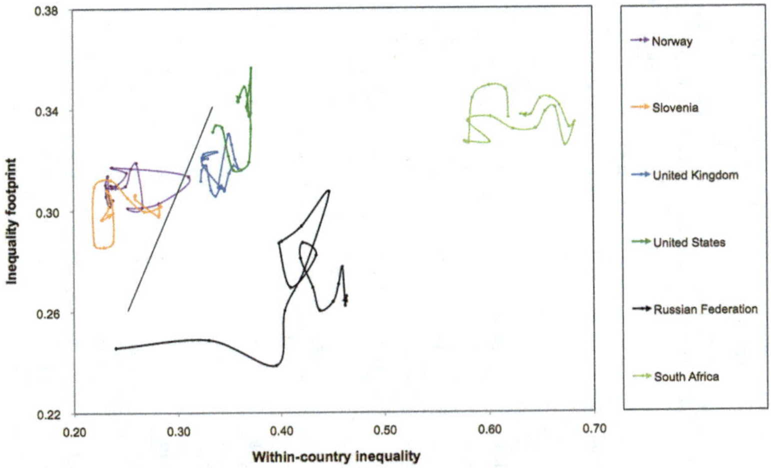

Fig. F.1 Time series of within-country Gini index

Other Examples

Table F.2 Additional information that supports Fig. F.2 (Latin America)

Year	Within country Gini index	Inequality footprint	Within country Gini index	Inequality footprint	Within country Gini index	Inequality footprint	Within country Gini index	Inequality footprint	Within country Gini index	Inequality footprint	Within country Gini index	Inequality footprint	Within country Gini index	Inequality footprint
	Argentina	Argentina	Bolivia	Bolivia	Chile	Chile	Ecuador	Ecuador	Colombia	Colombia	Mexico	Mexico	Brazil	Brazil
1990	0.47	0.45	0.42	0.42	0.55	0.38	0.54	0.38	0.51	0.37	0.50	0.33	0.61	0.31
1991	0.44	0.43	0.42	0.41	0.55	0.38	0.54	0.38	0.51	0.36	0.50	0.33	0.57	0.32
1992	0.45	0.41	0.48	0.40	0.55	0.37	0.54	0.37	0.51	0.36	0.50	0.33	0.54	0.32
1993	0.45	0.44	0.54	0.42	0.56	0.38	0.54	0.39	0.53	0.36	0.51	0.33	0.61	0.31
1994	0.46	0.44	0.55	0.42	0.55	0.39	0.54	0.39	0.54	0.36	0.52	0.33	0.61	0.31
1995	0.49	0.43	0.56	0.42	0.55	0.38	0.51	0.39	0.56	0.36	0.50	0.33	0.60	0.32
1996	0.50	0.44	0.57	0.43	0.55	0.38	0.51	0.40	0.57	0.37	0.49	0.33	0.61	0.33
1997	0.49	0.44	0.58	0.43	0.55	0.39	0.52	0.40	0.58	0.38	0.49	0.34	0.61	0.33
1998	0.51	0.44	0.58	0.44	0.56	0.40	0.52	0.41	0.58	0.39	0.49	0.35	0.60	0.34
1999	0.50	0.42	0.58	0.42	0.55	0.40	0.60	0.39	0.59	0.39	0.50	0.35	0.60	0.33
2000	0.51	0.41	0.63	0.42	0.55	0.39	0.57	0.40	0.59	0.39	0.52	0.35	0.60	0.35
2001	0.53	0.43	0.59	0.44	0.55	0.41	0.56	0.43	0.58	0.40	0.51	0.36	0.60	0.36
2002	0.54	0.44	0.60	0.43	0.55	0.41	0.56	0.44	0.61	0.40	0.50	0.36	0.59	0.36
2003	0.55	0.45	0.59	0.45	0.55	0.43	0.55	0.44	0.58	0.41	0.48	0.36	0.59	0.36
2004	0.50	0.44	0.59	0.43	0.54	0.41	0.55	0.44	0.58	0.40	0.46	0.35	0.58	0.34
2005	0.49	0.44	0.58	0.43	0.53	0.41	0.54	0.43	0.56	0.40	0.47	0.35	0.57	0.35
2006	0.48	0.44	0.56	0.43	0.52	0.41	0.53	0.44	0.59	0.40	0.48	0.36	0.57	0.35
2007	0.47	0.44	0.57	0.43	0.52	0.41	0.54	0.45	0.59	0.40	0.48	0.36	0.56	0.36
2008	0.46	0.42	0.56	0.42	0.52	0.40	0.51	0.43	0.57	0.39	0.48	0.35	0.55	0.36
2009	0.46	0.42	0.56	0.42	0.52	0.39	0.49	0.42	0.57	0.39	0.48	0.35	0.55	0.35
2010	0.44	0.42	0.56	0.42	0.52	0.39	0.49	0.42	0.56	0.39	0.48	0.35	0.55	0.35

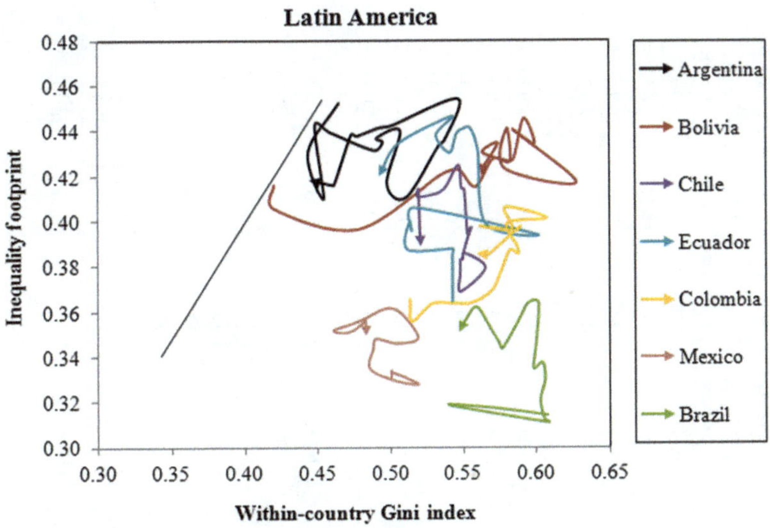

Fig. F.2 Time series of within-country Gini index

Table F.3 Additional information that supports Fig. F.3 (Europe)

Year	Within country Gini index	Inequality footprint	Within country Gini index	Inequality footprint	Within country Gini index	Inequality footprint	Within country Gini index	Inequality footprint	Within country Gini index	Inequality footprint	Within country Gini index	Inequality footprint
	Spain	Spain	France	France	Sweden	Sweden	Italy	Italy	Switzerland	Switzerland	Greece	Greece
1990	0.30	0.31	0.27	0.31	0.33	0.30	0.32	0.30	0.31	0.32	0.34	0.30
1991	0.32	0.31	0.28	0.32	0.32	0.30	0.32	0.30	0.31	0.32	0.34	0.30
1992	0.33	0.31	0.28	0.32	0.32	0.31	0.33	0.30	0.31	0.32	0.34	0.31
1993	0.34	0.31	0.28	0.31	0.32	0.30	0.33	0.30	0.30	0.32	0.34	0.30
1994	0.35	0.31	0.29	0.31	0.31	0.30	0.33	0.30	0.30	0.32	0.34	0.30
1995	0.35	0.31	0.28	0.31	0.30	0.30	0.34	0.31	0.29	0.33	0.34	0.30
1996	0.35	0.31	0.27	0.31	0.29	0.30	0.34	0.30	0.29	0.33	0.34	0.31
1997	0.34	0.32	0.26	0.31	0.28	0.30	0.35	0.31	0.28	0.33	0.34	0.31
1998	0.34	0.32	0.26	0.32	0.27	0.30	0.35	0.31	0.28	0.32	0.34	0.31
1999	0.34	0.32	0.27	0.31	0.26	0.30	0.36	0.31	0.28	0.32	0.34	0.30
2000	0.34	0.32	0.28	0.32	0.25	0.31	0.36	0.31	0.28	0.32	0.34	0.31
2001	0.33	0.33	0.27	0.32	0.24	0.31	0.35	0.32	0.28	0.33	0.35	0.32
2002	0.32	0.33	0.27	0.32	0.23	0.31	0.35	0.32	0.27	0.34	0.35	0.31
2003	0.31	0.31	0.27	0.32	0.23	0.29	0.34	0.30	0.28	0.32	0.36	0.30
2004	0.31	0.31	0.28	0.31	0.23	0.29	0.33	0.30	0.27	0.32	0.33	0.29
2005	0.32	0.32	0.28	0.32	0.23	0.29	0.33	0.30	0.31	0.32	0.33	0.30
2006	0.31	0.32	0.27	0.32	0.24	0.30	0.32	0.31	0.31	0.34	0.34	0.32
2007	0.31	0.32	0.27	0.32	0.23	0.30	0.32	0.32	0.32	0.35	0.34	0.32
2008	0.31	0.32	0.30	0.32	0.24	0.30	0.31	0.31	0.32	0.34	0.33	0.32
2009	0.32	0.32	0.30	0.32	0.25	0.30	0.32	0.31	0.30	0.33	0.33	0.31
2010	0.34	0.32	0.30	0.32	0.24	0.30	0.31	0.31	0.30	0.33	0.33	0.31

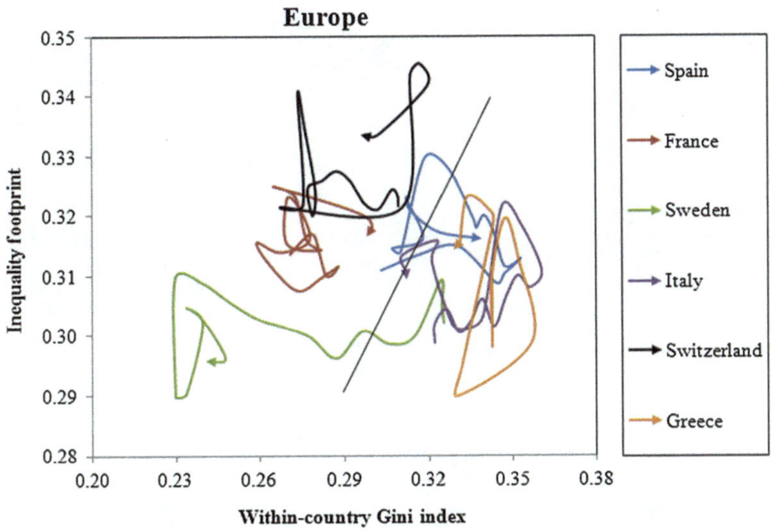

Fig. F.3 Time series of within-country Gini index

References

ABS. (2004). *Household expenditure survey—Detailed expenditure items*. ABS Catalogue No. 6530.0. Canberra, Australia: Australian Bureau of Statistics.

ABS. (2010). *Household expenditure survey—Detailed expenditure items*. ABS Catalogue No. 6530.0. Canberra, Australia: Australian Bureau of Statistics.

Bartelmus, P. (2007). SEEA-2003 accounting for sustainable development? *Ecological Economics, 61,* 613–616.

Costantini, V., Mazzanti, M., & Montini, A. (2011). *Hybrid economic-environmental accounts*. Oxfordshire: Taylor and Francis.

de Haan, M., & Keuning, S. J. (1996). Taking the environment into account: The NAMEA approach. *Review of Income and Wealth, 42,* 131–148.

Dietz, S., & Neumayer, E. (2007). Weak and strong sustainability in the SEEA: Concepts and measurement. *Ecological Economics, 61,* 617–626.

EC. (2001). *NAMEAs for air emissions: Results of pilot studies*. Luxembourg: European Communities.

Family Spending. (2010). Edition released: 30 November 2010. UK office for national statistics. Accessed January 25, 2013.

Family Spending. (2011). Edition released: 29 November 2011. UK office for national statistics. Accessed January 25, 2013.

Gallopin, G. (1997). Indicators and their use: Information for decision making. In B. Moldan & S. Billharz (Eds.), *Sustainability indicators (Report on the project on indicators of sustainable development)*. Chicheste: Wiley.

Horn, V. (1993). *Statistical indicators for the economic and social sciences*. Cambridge, UK: Cambridge University Press.

IBGE. (2010). *Pesquisa de Orçamento Familiares*. Rio de Janeiro, Brazil: Instituto Brasileiro de Geografia e Estatística, Ministério da Fazenda.

Ike, T. (1999). A Japanese NAMEA. *Structural Change and Economic Dynamics, 10,* 123–149.

Keuning, S. J., van Dalen, J., & de Haan, M. (1999). The Netherlands 'NAMEA'; presentation, usage and future extensions. *Structural Change and Economic Dynamics, 10,* 15–37.

Lange, G. M. (2007). Environmental accounting: Introducing the SEEA-2003. *Ecological Economics, 61,* 589–591.

Lenzen, M., Kanemoto, K., Moran, D., & Geschke, A. (2012). Mapping the structure of the world economy. *Environmental Science and Technology, 46,* 8374–8381. doi:10.1021/es300171x.

Lenzen, M., & Murray, J. (2011). Accounting for carbon flows: Comparing the principles of the UNFCCC and the SEEA. *Society and Natural Resources, 24,* 1216–1227.

OECD. (2009). Interpreting OECD social indicator in society at a glance 2009. OECD Social Indicators. OECD Publishing. http://dx.doi.org/10.1787/soc_glance-2008-4-en

Smith, R. (2006). Development of the SEEA 2003 and its implementation. *Ecological Economics, 61,* 592–599.

A. Alsamawi et al., *The Social Footprints of Global Trade*,
Environmental Footprints and Eco-design of Products and Processes,
DOI 10.1007/978-981-10-4137-2

Solt Frederick. (2009). Standardizing the world income inequality database. *Social Science Quarterly, 90*(2), 231–242 (SWIID Version 3.1, December 2011).
UNSD. (2003). Handbook of national accounting: Integrated environmental and economic accounting (SEEA 2003). New York, USA, http://unstats.un.org/unsd/envAccounting/seea2003.pdf, United Nations Statistics Division.
UNSD. (2009). System of national accounts 1993. Internet site unstats.un.org/unsd/sna1993/toctop.asp. New York, USA, United Nations Statistics Division.
Vaze, P. (1999). A NAMEA for the UK. *Structural Change and Economic Dynamics, 10,* 99–121.
Walker, B. H., & Pearson, L. (2007). A resilience perspective of the SEEA. *Ecological Economics, 61,* 708–715.

MIX
Papier | Fördert gute Waldnutzung
FSC® C083411

Kolibri 360 GmbH
Ferdinand-Jühlke-Straße 7
99095 Erfurt, Deutschland
produktsicherheit@kolibri360.de